HYPOTHERMIA

Title of related interest:

ADAM, J. M.   Hypothermia Ashore and Afloat

HYPOTHERMIA

MEDICAL AND SOCIAL ASPECTS

Proceedings of a symposium held at the
Manor House Hospital, Golders Green, London
8 November 1985

Edited by

DHARAM P. MAUDGAL

MBBS, PhD (Lon), MRCP (UK)

Consultant Physician/Gastroenterologist, Manor House Hospital,
London NW11
and
Associate Consultant Physician, St. George's Hospital, London SW17

PERGAMON PRESS

OXFORD · NEW YORK · BEIJING · FRANKFURT
SÃO PAULO · SYDNEY · TOKYO · TORONTO

U.K.	Pergamon Press, Headington Hill Hall, Oxford OX3 0BW, England
U.S.A.	Pergamon Press, Maxwell House, Fairview Park, Elmsford, New York 10523, U.S.A.
PEOPLE'S REPUBLIC OF CHINA	Pergamon Press, Qianmen Hotel, Beijing, People's Republic of China
FEDERAL REPUBLIC OF GERMANY	Pergamon Press, Hammerweg 6, D-6242 Kronberg, Federal Republic of Germany
BRAZIL	Pergamon Editora, Rua Eça de Queiros, 346, CEP 04011, São Paulo, Brazil
AUSTRALIA	Pergamon Press Australia, P.O. Box 544, Potts Point, N.S.W. 2011, Australia
JAPAN	Pergamon Press, 8th Floor, Matsuoka Central Building, 1-7-1 Nishishinjuku, Shinjuku-ku, Tokyo 160, Japan
CANADA	Pergamon Press Canada, Suite 104, 150 Consumers Road, Willowdale, Ontario M2J 1P9, Canada

First edition 1987

Library of Congress Cataloging in Publication Data

Hypothermia: medical and social aspects.
1. Hypothermia—Congresses.   2. Hypothermia—
Social aspects—Congresses.   I. Maudgal, D. P.
II. Manor House Hospital (London, England)
RC88.5.H97   1987   362.1'96988   86-25508

British Library Cataloguing in Publication Data

Hypothermia: medical and social aspects:
proceedings of a symposium held at the
Manor House Hospital, November 8th 1985.
1. Hypothermia   I. Maudgal, D. P.
616.9'88   RC88.5

ISBN 0-08-034188-8

Printed in Great Britain by A. Wheaton & Co. Ltd., Exeter

Foreword

THIS book results from the one-day symposium on the medical and social aspects of hypothermia which provided a forum for discussion among representatives from various professional bodies and voluntary organisations concerned with this topic.

Manor House Hospital which organised and hosted the event is a non-profit-making institution run by the Industrial Orthopaedic Society, registered under the Friendly Societies Act. Since 1919 the hospital has provided high-quality services to the society's nationwide membership, with a special emphasis on orthopaedic and general surgery. In recent years it has felt the need to extend its scope of treatment to include some medical and other services, a development symbolised by this symposium with its emphasis on illness prevention among those most at risk — the elderly.

The first section concentrates on the medical and physiological aspects of hypothermia discussed by a group of eminent doctors from London teaching hospitals where they are engaged in active research and provision of day-to-day care to the elderly with acute illnesses. Any discussion of hypothermia would not be complete without mentioning its social aspects. The second section is devoted to discussion of the social problems faced by the elderly which contribute to cold-related illnesses. The role of statutory social services and of voluntary organisations in preventing this potentially fatal condition is shown to be complementary to the work of health care personnel at both primary and hospital levels.

D. P. MAUDGAL

Contents

List of Contributors

P. BARON
Field Assistant to Director General
Help the Aged
St James Walk
London EC1

G. C. J. BENNETT MB, MRCP
Consultant Physician in Geriatric Medicine
The London Hospital (Mile End)
London E1

G. P. J. BEYNON MA, MRCP
Consultant Geriatrician
The Middlesex Hospital
Mortimer Street
London W1

A. T. BRAIN MA, FRCP
Consultant Physician in Geriatric Medicine
City and Hackney Hospital
Homerton High Street
London E9

K. J. COLLINS BSc, MB, DPhil, MRCP
Department of Geriatrics
St Pancras Hospital
St Pancras Way
London NW1 0PE

M. COLLYER
Health Needs Development Officer
Age Concern, Greater London
Knatchbull Road
London SE5 9QY

N. F. COSIN
Deputy Area Group Head
London Borough of Camden
Department of Social Services
West End Lane
London NW6

A. GRIMES
Assistant Director (communications)
Scottish Council for Community and Voluntary Organisations
Claremont Crescent
Edinburgh EH7

M. IMPALLOMENI MD, FRCP
Consultant Geriatrician/Senior Lecturer in Medicine
Royal Postgraduate Medical School
Hammersmith Hospital
London W12 0HS

J. JONES CH
President
Retired Members Association
Transport & General Workers Union
Transport House
Smith Square
Westminster
London SW1

W. R. KEATINGE MB, PhD
Professor of Physiology
The London Hospital Medical College
Turner Street
London E1 2AD

D. P. MAUDGAL MB, PhD, MRCP
Consultant Physician/Gastroenterologist
Manor House Hospital
Golders Green,
London NW11

Overview

D. P. MAUDGAL

Manor House Hospital, London NW11

EACH year in England and Wales during the Winter months over 40,000 more people die than in summer. The number is still higher if the weather is colder. This enormous increase in the number of deaths during winter months is predominantly due to cold-related diseases. The true number of deaths due to hypothermia, a condition where central body temperature falls below 35°C (95°F), is difficult to ascertain, but is thought to account for only about 1 per cent of deaths during the winter months. Hypothermia as a special category in mortality statistics was introduced as late as 1979. In that year the number of deaths certified due to hypothermia was 429. Due to increased awareness and steps taken to minimise the risks of cold, the deaths certified due to hypothermia has fallen to 219 by 1983. This decrease in death rate is encouraging, but leaves a lot more to be achieved. Since the diagnosis of hypothermia can be easily missed, especially in those patients who present with other associated illnesses, the true number of deaths due to hypothermia is probably much higher than recorded.

Those most likely to fall victim to hypothermia are the elderly, the disabled and those on very low incomes. Elderly people constitute by far the largest group suffering from hypothermia during the winter. A number of factors are responsible for making them susceptible to hypothermia.

The process of ageing is associated with a number of physiological changes in the body. One such change is the inability of the body to maintain its internal temperature. In a younger person the body's temperature is maintained by a balance between the production and the loss of heat from the body. This is controlled by the thermal regulatory centre situated in the part of the brain called the hypothalamus. All three aspects of temperature control, i.e. heat production, thermoregulatory centre and heat loss, are likely to be affected in the elderly. The basic metabolic rate in the elderly is lower than in the young, causing reduction in the production of metabolic heat which is further reduced by lack of physical activities due to associated illnesses. Elderly people tend to lose more heat because of their high surface area:body mass ratio. In this respect they tend to resemble the very young. A number of elderly people also have defective temperature perception so that they fail to recognise changes in their environmental temperature until it has fallen to very low levels. Their ability to make necessary adjustments is thus dangerously impaired.

Unfortunately the process of ageing is also associated with onset of a number of other illnesses. These illnesses can contribute to hypothermia by reducing heat production, e.g. hypothyroidism, hypopituitarism, diabetes mellitus and malnutrition; by reduced activity, e.g. arthritis, parkinsonism and the after-effects of strokes; by increased heat loss, e.g. by increased dilatation of the superficial blood vessels as seen in extensive psoriasis, eczema, exfoliative dermatitis and paget's disease of the bones, especially when involving the skull. A number of drugs prescribed to treat the illnesses mentioned above can contribute to defective thermoregulation.

In addition to disturbed body functions due to ageing or other associated illnesses and drugs, a number of social factors play an important part in bringing about hypothermia in the elderly. A survey carried out by Vicks in 1971–2 demonstrated that 75 per cent of old-age pensioners maintain their living room temperatures below the DHSS recommended living room temperatures of 65°F. There is some evidence to suggest that the number of admissions to hospital tends to increase a few days after a spell of cold weather. These admissions are due not only to hypothermia but also to a number of other illnesses initiated, or made worse, by the cold weather. The main reasons for the elderly continuing to live in cold surroundings are:

(a) Low income, which tends to decrease further as one advances in age from mid 60s to the over 70s.
(b) High cost of fuel, which has increased manifold during the last decade, puts an extra strain on the meagre income of the elderly.
(c) Elderly people tend to live in older property which is more costly to maintain and heat.
(d) In older properties heating appliances and insulation are not satisfactory, which leads to a high percentage of wasted energy.

The onset of hypothermia is associated with disturbed cerebral and cardiac functions. The person becomes listless and confused and does not usually remember events at the time of low body temperature. Unconsciousness sets in at very low body temperatures. Hypothermia also affects respiration, kidneys and liver. Unless a low-reading thermometer is used to take the temperature, diagnosis of hypothermia may be missed and the patient may be treated for the manifestations of hypothermia instead of hypothermia itself. Once the diagnosis of hypothermia has been established, the cornerstone of treatment is slow rewarming of the body using a well-insulated sleeping bag or a space blanket, depending upon availability. Symptoms from associated heart, lung, kidney and liver disorders need appropriate monitoring and treatment. In special centres rapid rewarming of the body may be undertaken, but this should not be practised outside these centres.

The most appropriate treatment of hypothermia is its prevention. A number of steps can be taken by relatives or neighbours which will help to avoid the onset of hypothermia in an elderly person. These are:

1. *Diet.* A malnourished person can develop hypothermia even in warm temperatures, but is especially at risk of developing hypothermia in the winter. An elderly person should be encouraged to take a proper diet containing an adequate amount of protein. Attention should be paid to ensuring that all elderly people have at least one hot meal per day and plenty of hot drinks. This will provide fuel to produce internal heat.
2. *Clothing.* It is very important to wear proper warm clothing, particularly around the extremities of the body (the hands, feet and head), in order to minimise heat loss.

3. *Bedroom temperature*. The bedroom is the part of the house which often remains cold all day and night. A person usually spends a third of his or her life in the bedroom. Provision of warm bedclothing and an electric thermal blanket, along with some heating in the bedroom if possible, should be achieved. If it is not possible, the bed should be moved in to the living room where it may be warmer.

4. Efficient and reliable heating of the house is vital. The recent sharp rise in the cost of fuels has put extra strain on the meagre resources of the elderly. There are, however, additional benefits available from the DHSS for those on supplementary benefit. Such grants include heating and clothing allowances, attendance allowance which allows an immobile person to pay for home helps, etc.

5. Elderly people should be encouraged to be as mobile as possible. Every attempt should be made to involve them in outdoor activities.

Section I

HYPOTHERMIA

MEDICAL ASPECTS

Hazards of Cold Weather

W. R. KEATINGE

The London Hospital Medical College, Turner Street, London E1 2AD

HYPOTHERMIA was first recorded as a special category in the mortality statistics in 1979. In that year, records for England and Wales showed 429 deaths as due to hypothermia. Steps taken since then have been associated with almost a halving of these numbers, to 219 in 1983, in spite of a steadily increasing population of elderly people at risk. This reflects credit on everyone concerned, but our job as scientists is also to stand back and assess how much hypothermia contributes to preventable illness and death, in relation to other causes of serious illness, in elderly people in winter.

There is an overall excess mortality of about 40,000 deaths in winter every year in England and Wales. This has often been associated with hypothermia, simple cooling of the body core until death occurs, but the recorded deaths from hypothermia account for less than 1 per cent of these excess deaths. About half of the excess deaths were due to heart attacks and strokes, as recorded on death certificates.

Early surveys of people's mouth temperatures suggested that hypothermia might be quite common, and that more people may have died of hypothermia than the statistics record. However, mouth temperature gives false low readings when people are in cold surroundings and the face is cold, and measurements using more reliable measures of deep body temperature indicate that seriously low body core temperature is not common in elderly people at home. The numbers of patients admitted to hospital with hypothermia have also been small, and those patients who

3

were hypothermic were usually cold only because drink or drugs or serious illness has caused them to collapse in cold surroundings. However, this still left some possibility that unrecognised hypothermia was the real cause of some of the deaths recorded as due to heart attacks and strokes, and support for this possibility came from the fact that there was no obvious reason why heart attacks and strokes should be more frequent in winter time. Studies of the clotting factors in blood had not shown any seasonal changes of a kind that could explain the recorded increases in heart attacks and strokes. Such deaths are largely due to arterial thrombosis. We therefore made some experiments on volunteers to look for any change that might explain an increased clotting of blood in arteries when people are exposed to cold. These showed a series of effects of cold which are harmless in themselves, but over sufficient time would increase the risk to people who are already liable to an early thrombosis.

The cooling in these experiments was of a degree that would often occur in winter, and was brought about by cool air with a strong wind that lowered skin temperature near to 24°C while rectal temperature fell little, only 0.6°C compared to control experiments. Apart from increases in arterial blood pressure, and in the concentration of red cells in the blood, this caused increases in the number of platelets per unit volume of blood. We will be hearing later some evidence that the increases in arterial blood pressure in the cold are at least as large in elderly people as in younger adults. All of these changes, if they are sustained long enough, could increase the risk of a variety of circulatory problems, and together they seem able to account for the increased cardiovascular hazards in winter. The main practical point is that it is the ordinary, everyday degree of exposure to cold that people encounter in winter that seems to cause most of the problems, rather than severe cold and hypothermia.

What of the remaining excess deaths in winter that are not due to cardiovascular causes? Some can clearly be explained by obvious factors such as people falling on slippery roads in freezing weather and having accidents in vehicles in fog and ice. A few are due to hypothermia. Accidents indoors also increase, and it is possible that borderline degrees of body cooling may contribute to these. There are also excess deaths due to increased respiratory infections, which may be due both to altered patterns of life in winter and to direct effects of the breathing of cold air on the respiratory passages.

As regards prevention, our evidence about the causes of winter mortality therefore presents a less easy problem than seemed to exist until recently. It is likely that the work that has been done in improving house heating and insulation, and home care of the elderly, in recent years has improved the comfort of elderly people and reduced those cases of hypothermia that do occur. It is not at all clear that it will prevent the other causes of excess death in winter, and excessive concentration on hypothermia could be counterproductive. The main thing we need for future progress is more information about the precise types of cold exposure and pattern of life that cause these problems. It is likely that if all elderly people lived indoors in a constant thermally controlled environment for 24 hours a day, winter mortality would cease. Apart from practical problems, this is not likely to provide a reasonable quality of life in a country where outdoor activities and mobility are so important. Retirement to warm climates could be a better and an economically more practical solution if stable international arrangements were made to make this easier. However, we do not yet know in any detail the particular elements of winter life that cause the hazards to elderly people, and it is likely that much could be done by advice on patterns of life and buildings to reduce winter hazards to the elderly in this country, without harming the quality of life. The main object must be to reduce the incidence of crippling and fatal illness in those elderly people in which quality of life and expectancy of life is otherwise good. The main need is for more information on the factors responsible for them. In the meantime, the practical message must be that any exposure to cold may carry some risk to the fit elderly and that they should try to minimise such exposure without seriously interfering with their outdoor activities and recreations.

REFERENCES

Bull, G. M. and Morton, J. (1978) Environment, temperature and death rates. *Age Ageing* 7, 210–224.

Keatinge, W. R., Coleshaw, S. R. K., Cotter, F., Mattock, M., Murphy, M. and Chelliah, R. (1984) Increases in platelet and red cell counts, blood viscosity, and arterial pressure during mild surface cooling: factors in mortality from coronary and cerebral thrombosis in winter. *Brit. Med. J.* 289, 1405–1408.

Physiological changes in the elderly predisposing to hypothermia

K. J. COLLINS

Department of Geriatric Medicine, School of Medicine, University College London, St Pancras Hospital, St Pancras Way, London NW1 OPE

INTRODUCTION

A characteristic of the functional changes which accompany ageing is the inability to respond efficiently to environmental stresses and to adapt to these stresses. One important example of loss of adaptation is the poor response of many elderly people to heat and cold. The elderly, like other at-risk groups such as young children, the disabled and sick, are vulnerable to ambient temperature change and may develop hypothermia more readily in severely cold conditions.

Statistics available for the UK population, however, do not support the view that there are large numbers of elderly people suffering from clinical hypothermia[1]* though there may be a larger number in whom hypothermia is undiagnosed when the condition occurs secondary to other disorders. The only real evidence we have of the magnitude of the problem is from mortality statistics, and these usually include figures for both elderly and young people. There is the likelihood that hypothermia cases and deaths increase in Britain when winter conditions become more severe (Table 1). Of greater significance for the elderly is the effect of low ambient temperatures on morbidity and mortality from respiratory and cardiovas-

---

*Superscript numbers refer to references at end of chapter.

TABLE 1. Mean Outdoor Temperatures and Hypothermic Deaths in England and Wales

Year	Mean monthly dry bulb temperature (°C)			Hypothermia deaths for Jan. — Mar. quarter (hypothermia as underlying) cause)
	Jan.	Feb.	Mar.	
1979	0.9	1.9	5.3	293
1980	3.1	6.3	5.4	137
1981	5.2	3.7	8.3	151
1982	3.6	5.5	6.6	225
1983	7.0	2.6	7.0	121
1984	4.3	4.2	5.3	145
1985	1.0	2.3	4.6	277

Sources: Office of Population Censuses and Surveys; London Meteorological Office, London

cular diseases. For example, blood pressure may rise significantly in ambient temperatures below 12°C dry bulb[2] and increase the risk of coronary or cerebrovascular accidents. In recently published statistics on winter mortality[3] it is postulated that for every degree Centrigrade fall in the average winter temperature (December–March) there is a rise in the number of winter deaths by about 8000. A very small proportion of these excess winter deaths appear to be associated with certification of clinical hypothermia.

PREDISPOSING FACTORS

Hypothermia in the elderly may develop because of a combination of both extrinsic and intrinsic factors (Table 2). Many of these factors are present together, and this increases the risk of cold-induced illness in the elderly group.

In a survey carried out by the Office of Population Censuses and Surveys in 1976 it was found that 90 per cent of elderly people over 65 years of age in England lived in their own or own families' housing, and only 8 per cent lived in sheltered accommodation or old peoples' homes.[4]

TABLE 2. Hypothermia in the Elderly: Predisposing Factors

Extrinsic Factors
  1. Cold outdoor and indoor temperatures
  2. Poor domestic heating and insulation
  3. Lack of warm clothing
  4. Inadequate nutrition

Intrinsic Factors
  1. Low metabolic heat production
  2. Immobility
  3. High surface area:mass ratio in frail elderly
  4. Many clinical conditions leading to secondary hypothermia:
      e.g. Hypothyroidism, hypopituitarism, diabetes mellitus
           Atherosclerosis, myocardial infarction, stroke
           Parkinsonism, paraplegia
           Mental impairment, confusional states
           Acute and chronic infections, alcoholism

  5. Drugs
      e.g. Phenothiazines, hypnotics (especially in excess)
           Hypoglycaemic agents
  6. Slow behavioural responses to cold
  7. Impairment of thermal perception
  8. Less efficient thermoregulation

A high proportion of elderly people, therefore, have at least some control over their own indoor environment. As will be shown, elderly people often show poor behavioural responses to temperature changes and they may not be efficient at controlling indoor temperature conditions. Other old people have a blunted sense of cold and cold discomfort and may think it safe to compromise their heating requirements. The worst combination of extrinsic and intrinsic factors occurs when old people with less efficient body temperature regulation live in the oldest dwellings and are less able to maintain their properties or to pay for repairs, heating or house insulation. Of the physiological factors operating to reduce the efficiency of elderly people to cope with cold conditions, four main aspects will be considered here. They are (a) low metabolic heat production, (b) impairment of thermal perception, (c) slow behavioural responses to cold, and (d) less efficient thermoregulation.

Metabolic Heat Production

The proportion of body mass made up of actively functioning cells, as distinct from body fluids, fat stores and structural elements, is usually smaller in the elderly. This results in an overall decrease in total metabolic heat production whether this is measured as heat produced per kilogram of body weight or as heat produced per square metre of body surface. The normal changes in basal metabolic rate with age are given in Table 3. Apart from this age-related decline in resting metabolic rate, the more

TABLE 3. Basal Metabolic Rate in Relation to Age

Age (years)	BMR (watts per square metre)	
	Males	Females
2	64	61
10	55	53
20	48	42
30	45	41
40	44	41
50	43	40
60	42	37
70	40	37

Source: The Mayo Foundation

sedentary life-style of the elderly results in a smaller contribution to overall heat production from muscular activity, while a smaller energy intake decreases the amount of extra heat produced by the calorigenic action of food.

Normal body temperatures can be maintained if a lower body heat production is compensated by a reduced rate of heat loss. In frail, old people there is a relative increase in heat loss from the body surface due to a high surface area:body mass ratio and sometimes an inability to constrict the blood vessels of the skin sufficiently to improve surface insulation. So not only is there a smaller heat production but a tendency for greater surface heat loss. Body temperature does not fall, however, for there are yet other compensatory mechanisms which help prevent this. Behavioural thermoregulation usually ensures that frail elderly people are well clothed and are not exposed unduly to cold. Lower metabolic heat production may also be partly balanced by less insensible evaporative water loss from the skin.[5] The problem arises when compensatory mechanisms begin to fail and it is in this situation that the deep body temperature can no longer be maintained in cold surroundings.

One of the compensating mechanisms for increasing heat production is shivering. The suggestion has been made that shivering thermogenesis is reduced or even absent in the elderly. Our own investigations imply that shivering ability is not completely lost with ageing, even in those over 80 years, but changes occur in the character of the response.[6] High peaks of muscle contraction achieved by young people are not usually attained by the elderly, and there is often a longer latent period required to initiate maximum shivering. Ageing changes in the fast and slow motor units in skeletal muscle itself, the lack of muscle bulk and lack of muscle training may all account for these differences in shivering thermogenesis.

In the adult human, non-shivering thermogenesis from brown fat does not seem to play a major role in thermoregulation. Any contribution to internal heat production from brown fat in later years is likely to be small, for although a few cells of brown fat have been found in tissues of younger adults, they disappear almost entirely by the eighth decade of life.

Thermal Perception

Information about the temperature of the environment is conveyed from thermoreceptors in the skin to the brain where it is interpreted (a) as a temperature sensation of warmth or cold, (b) to give an impression of thermal comfort whether pleasant or unpleasant, and (c) to bring about involuntary thermoregulatory responses such as shivering or vasoconstriction. The function of the thermoreceptors, like other sensory systems, may alter in old age. It is known that cold receptors in the skin of some animals are highly dependent for optimum function on a good blood supply, but in ageing skin the vascular supply may become reduced.

In tests designed to measure the ability to discriminate between the temperature of objects it has been found that whereas young people could perceive a difference of 1°C between two objects, older people often could not match this (Table 4); sometimes differences as great as 3°C could not be detected. One explanation for this may be that older people are less confident in reporting differences in temperature rather than being less capable of detecting them. However, by using signal detection analysis it is possible to estimate the way such decisions are made by individuals. In these studies there appeared to be no differences in the criteria upon which decisions were taken, and this suggests that the temperature perception differences may have been a true age effect. One study of elderly people aged between 74 and 86 years showed that most perceived that indoor conditions were cold and responded in an appropriate way by increasing indoor heating. Some, however, experienced the sensation of cold only at unusually low temperatures.[7]

The perception of ambient temperature also contributes an important component to the overall feeling of thermal comfort. For most normal healthy people the range of climates which young and elderly consider to be comfortable is the same,[5] given that clothing and physical activity are standardised (Table 4). If a group of elderly people are selected on the basis of abnormally poor temperature discrimination, then it is found that they may also have an abnormal comfort sensation and feel more comfortable in colder environments than most.[8]

Behavioural Thermoregulation

Since it is important to determine optimum ambient temperatures in the homes of the elderly, investigations have been made to determine how temperature preferences are made by the individual. Young and elderly volunteers were asked to sit alone in a rapidly responding temperature-controlled room for a period of 3 hours.[9] After 30 minutes during which time the room was kept at a neutral 19°C, control of the heating and cooling was taken over by the volunteer. By operating a switch the room could be temperature-controlled to suit the comfort of the individual. Some of the elderly performed this task as efficiently as the young group. Others, however, controlled room temperature very poorly, allowing the temperature to change wildly instead of gradually narrowing the temperature swings down to the preferred comfort level. The temperature-control curves were analysed to give the range between the highest and lowest

TABLE 4. The Behavioural Response of Healthy Young and Elderly Subjects in the Control of Indoor Temperature. The subjects were sedentary and wearing 0.8 clo.**
(Means ± SD)

	Elderly ($n = 17$)	Young ($n = 13$)
Age (years)	73.8 ± 2.7	26.7 ± 7.2
Preferred ambient temperature (°C)	22.1 ± 2.5	22.7 ± 1.2
Temperature discrimination (°C)	1.9 ± 0.6	0.8 ± 0.4*
Maximum ambient temperature range of control (°C)	8.2 ± 3.9	4.8 ± 1.6*
Final ambient temperature range of control (°C)	4.4 ± 3.7	2.4 ± 1.7*
Frequency of ambient temperature change (no./hr.)	2.8 ± 1.6	5.6 ± 2.4*

*$p< 0.001$.

**0.8 clo = light indoor clothing, i.e. shirtsleeves.

ambient temperatures during the 3 hours, the final range of temperatures in the last 30 minutes and the frequency of temperature changes (Table 4). The lack of precision of many of the elderly in adjusting the room temperature suggests that this may contribute to their vulnerability in cold conditions.

The Thermoregulatory System

Deep body temperature is controlled by the integrated action of a sensory system relaying thermal information from thermoreceptors in the skin and other deep structures such as muscles, the controlling centres in the hypothalamus of the brain, the efferent nerves from the centres and the effector organs which evoke the physiological responses to regulate temperature. Ageing may affect the efficiency of any one or combination of these components. Some evidence for changes in shivering capacity have already been described. In the zone of vasomotor control between the onset of shivering or sweating, vasoconstriction or vasodilatation of the skin blood vessels forms the major means of defence against temperature stress. In the cold, there have been several investigations that have demonstrated abnormal vasoconstrictor patterns in elderly people.[9,10] Some old people, perhaps 20 per cent of a normal healthy group, do not vasoconstrict as fully and are therefore in danger of losing more heat than normal in the cold. In neutral conditions young people also possess a vasoconstrictor rhythm, i.e. transient bursts of vasoconstrictor activity occurring 2 or 3 times a minute. This rhythm appears to be generated in the central nervous system, because electrical recordings from nerves to the blood vessels show a similar rhythm. In many elderly people this rythmic activity is absent, and it suggests an altered sensitivity of the vasomotor control system.

Recent work aimed at studying the responsiveness of the blood vessels themselves[11] shows that the compliance of both the venous and arterial blood vessels decreases with age and that reduced arterial compliance in the cold may significantly affect the vasoconstrictor response to cold in elderly patients with arteriosclerosis.

The temperature-control centres in the brain are also likely to be involved in the ageing process. Changes in the integrity of the vital hypothalamic centres due to degenerative diseases in old age can result in hypothermia or hyperthermia. Ageing appears to reduce the fine control of temperature regulation with resetting or desynchronisation of the central nervous control which appears to allow the body temperature to oscillate uncontrolled between wider limits of internal temperature before physiological adjustments are made.

CONCLUSIONS

The efficiency of thermoregulation is reduced in a proportion of old people. The underlying cause for this is that structural and functional changes take place in the nervous system and the blood supply to organs and tissues is reduced during ageing. With a higher incidence of degenerative disease, decreased cardiovascular capacity (often due to detraining) and a higher incidence of mild confusional states, there is a tendency for the elderly to develop a non-responsive attitude to the threat of a cold environment. As a result there is a greater dependency of the old on their physiological mechanisms of temperature regulation. However, the investigations summarised here suggest that physiological capacity is reduced: internal heat production is often smaller, thermal perception may be blunted, behavioural thermoregulation is less precise and the temperature-regulating responses of the body become less efficient. To summarise, thermoregulation in the elderly does not fail, but the potential to respond efficiently is diminished and this leads to an increased vulnerability to cold and the threat of hypothermia.

REFERENCES

1. K. J. COLLINS, (1983) *Hypothermia: the Facts.* Oxford: Oxford University Press.
2. K. J. COLLINS, EASTON, J. C., BELFIELD-SMITH, H., EXTON-SMITH, A. N. AND PLUCK, R. A. (1985) Effects of age on body temperature and blood pressure in cold environments. *Clinical Science,* **69**: 465–470.
3. M. R. ALDERSON (1985) Season and mortality. *Health Trends,* **17**: 87–96.
4. A. HUNT (1978) *The Elderly at Home.* OPCS Social Services Division. London: HMSO.
5. P. O. FANGER, (1972) *Thermal Comfort.* New York: McGraw-Hill.
6. K. J. COLLINS and EXTON-SMITH, A. N. (1983) Thermal homeostasis in old age. *Journal of the American Geriatrics Society,* **31**: 519–524.
7. A. J. WATTS (1972) Hypothermia in the aged: a study of the role of cold sensitivity. *Environmental Research,* **5**: 119–126.
8. K. J. COLLINS, EASTON, J. C. AND EXTON-SMITH, A. N. (1982) The ageing nervous system: impairment of thermoregulation. In: *Advanced Medicine* No. 18 pp. 250–257. Royal College of Physicians, London, Ed. M. Sarner. London: Pitman.
9. K. J. COLLINS, EXTON-SMITH, A. N. and DORÉ, C. (1981) Urban hypothermia: preferred temperature and thermal perception in old age. *British Medical Journal,* **282**, 175–177.
10. S. M. HORVATH, RADCLIFFE, C. E., HUTT, B. K. AND SPURR, G. B. (1955) Metabolic responses of old people to a cold environment. *Journal of Applied Physiology,* **8**: 145–148.
11. K. J. COLLINS, DURNIN, C. J. A. AND PLUCK, R. A. (1986) Peripheral venous compliance and vasomotor responses to cooling and warming in young and elderly subjects. *Journal of Physiology* (in press).

Metabolic and Other Causes of Hypothermia

M. IMPALLOMENI

Royal Postgraduate Medical School, Hammersmith Hospital, Du Cane Road, London W12 0HS

HYPOTHERMIA, although rare, is most commonly seen at the extremes of life,[1]* in children and in the very old. It is often encountered during winter months, but can occur at all times of the year.

For a rational discussion of this intriguing condition, I propose to follow two classifications summarised in Tables 1 and 2.

TABLE 1.   Hypothermia in Old Age

1.	Primary hypothermia:	due to a primary defect of the temperature control system, or thermostat.
2.	Secondary hypothermia:	low body temperature is a symptom of another disease.
3.	Exposure, or accidental hypothermia:	hypothermia is the result of exposure to a cold environment.

In real life, however, things are not so simple and tidy. Most elderly people admitted to hospital with hypothermia were living alone, fell in the night and spent several hours on the floor in their cold house.

Thermoregulation in old age does not fail, but is less efficient, especially in the sick and very old. It is prone to be deranged by diseases or drugs, especially in a cold environment, and result in hypothermia.[2]

---

*Superscript numbers refer to references at end of chapter.

TABLE 2.   Secondary Hypothermia

---

1.   Widespread, generalised fall in metabolic rate and internal heat production:
     e.g. hypothyroidism
          diabetes mellitus
          hypopituitarism
          malnutrition

2.   Decreased heat production by the muscles as a result of decreasing mobility:
     e.g. stroke
          severe arthritis
          Parkinson's disease

3.   Circulatory problems:
     (a)   Poor circulation:
           e.g. coronary thrombosis
                severe heart failure
                overwhelming bacterial infection
     (b)   Increased peripheral circulation:
           e.g. erythroderma
                Paget's disease of bones

4.   Diseases of the central or peripheral nervous system:
     e.g. stroke
          mental illness
          brain tumors
          spinal cord and peripheral nerves disease

5.   Drugs:
     Especially those acting on the central nervous system. They may impair the subject's
     awareness of cold, and also impair temperature control.

One should always be on the lookout for such disorders, as the patient's prognosis depends on the diseases causing or precipitating it, or associated with hypothermia, which has a morbidity of its own.

I will discuss and illustrate a few examples of the conditions referred to under secondary hypothermia.

DECREASED METABOLIC RATE AND INTERNAL HEAT PRODUCTION

The patient shown in Fig. 1 was admitted to hospital with hypothermia. He was found to be *myxoedematous*. He responded well to therapy and is now living a normal life years after this episode. This is, however, rare: in my experience less than 5 per cent of hypothermia patients in old age are

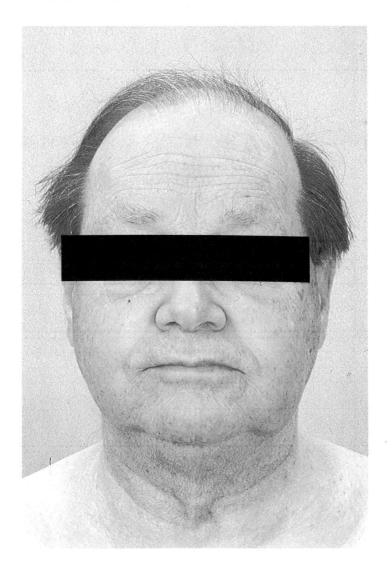

Fig. 1a.   67-year-old man soon after recovering from hypothermia due to myxoedema.

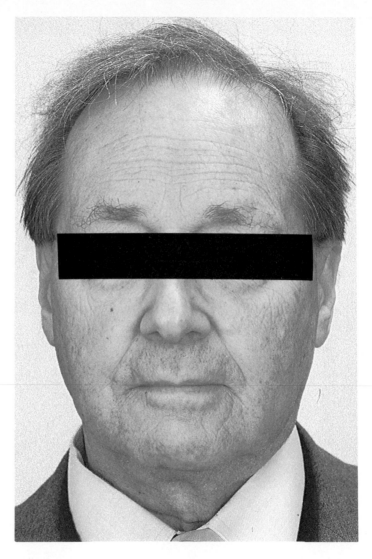

Fig. 1b.   The same man 3 months later.

found to be hypothyroid. In a series of 13 elderly patients with myxoedema coma,[3] only 3 out of 13 presented with hypothermia. Severe hypothyroidism can be easily mimicked by hypothermia, where the face may appear white and puffy, the senses dulled, the pulse slow and weak, and the tendon reflexes sluggish. Characteristically, in severe myxoedema only the relaxation phase of the reflexes is delayed, whereas in hypothermia both the contraction and relaxation phases are prolonged.

Pituitary Disease

Frank pituitary disease is not common in old age. I have encountered only one patient whose hypothermia (29.8°C) was due to pituitary failure. This was a 66-year-old woman who had suffered a severe gastrointestinal tract haemorrhage when in her teens, and never had any menses. After rewarming, her blood glucose was found to be 0.8 mmol/l, her T4 much decreased. On adequate hormonal replacement therapy she made a full recovery.

Diabetes Mellitus

This is common in the elderly. The disorder of sugar metabolism may be so severe as to decrease production of heat significantly, as in keto-acidotic or hyperosmolar non-ketotic crisis, but this is rare.

Drugs used in its treatment may also precipitate hypothermia: insulin and most oral antidiabetic drugs. Diabetes is also often complicated by disease of the central and peripheral nervous system, which may impair temperature control.

Malnutrition

This is probably more common in the elderly than formerly thought (Fig. 2). It may be so severe that the body just does not have enough fuel to keep warm, with a consequent drop in heat production, but this is rare.

The subcutaneous tissues may be so thin that the body does not have enough insulation, with consequent heat loss. Malnourished people are

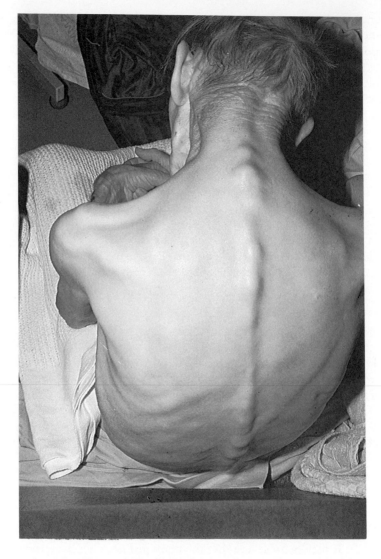

Fig. 2a,b. Severe malnutrition.

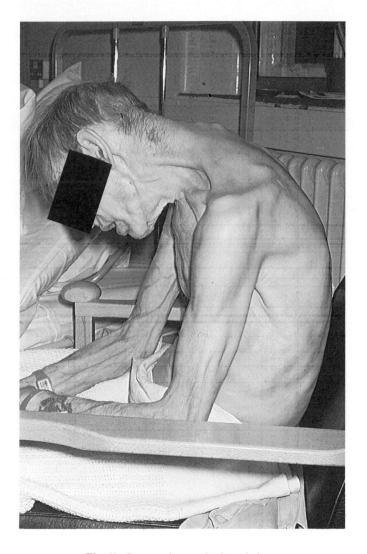

Fig. 2b. Same patient as 2a, lateral view.

also often deficient in some important vitamins which are necessary for the normal functioning of the complex thermoregulatory mechanisms.

INACTIVITY

Any condition which restricts physical activity, such as stroke, extensive arthritis, Parkinson's disease, spinal cord and peripheral nerves disease. These patients have a decreased production of heat and run the danger of falling on the floor, and of being unable to get up for hours. In some published series up to 65 per cent of hypothermic elderly patients were found on the floor in their homes where they had been for several hours.

Figures 3 and 4 show two patients, both admitted to hospital with hypothermia. The first shows a patient with severe rheumatoid arthritis, affecting most synovial joints in the body and with bilateral painless dislocation of the elbow joints. The second, severe degenerative arthritis of the knee joints with effusion, and gross wasting of the thigh muscles.

POOR CIRCULATION

Coronary thrombosis or severe pulmonary emoblism may be followed by a severe drop in cardiac output with decreased perfusion of most organ systems in the body, which may impair temperature control and result in hypothermia. This is, however, rare. Hypothermia may also occur in overwhelming bacterial infections, such as gramnegative septicaemia, although its pathogenesis here is not well understood.

Increased blood flow through abnormally dilated blood vessels in the skin or in the skeleton may also be associated with hypothermia. Generalised erythroderma (Fig. 5), as in extensive psoriasis, eczema, exfoliative dermatitis, interferes with heat preservation by the body, which loses heat like a radiator turned on all the time. In one series of elderly hypothermic patients reported in the literature[4] 10 per cent of all patients suffered from this condition. I have observed one patient with erythroderma develop hypothermia during the summer in a hospital bed.

Figure 6 illustrates a patient with severe Paget's disease of bone. This may cause hypothermia if a large part of the skeleton is affected, especially the skull.

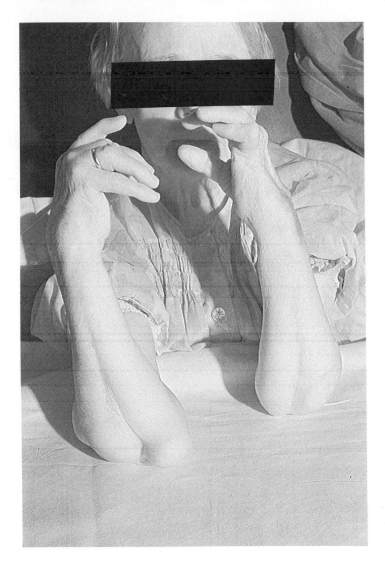

Fig. 3.   Severe rheumatoid arthritis with dislocated elbow joints.

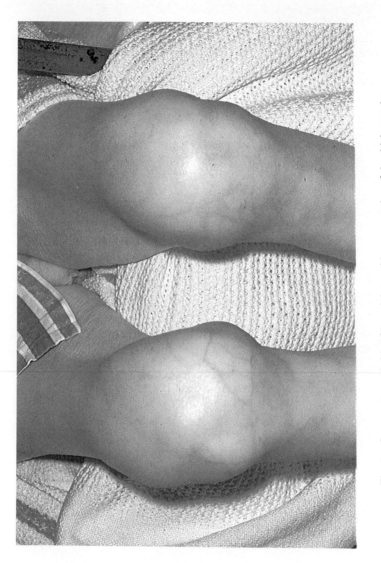

Fig. 4. Severe osteo-arthritis of knee joints, with gross wasting of the thigh muscles.

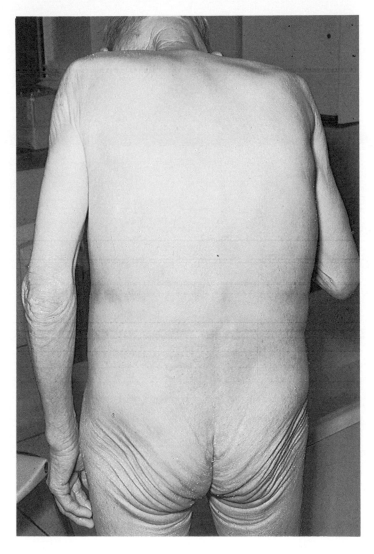

Fig. 5. Erythroderma due to severe eczema.

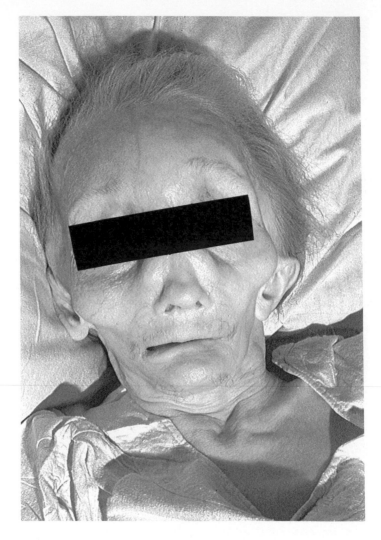

Fig. 6. Paget's disease with gross deformity of the skull.

DISEASES OF THE CENTRAL AND
PERIPHERAL NERVOUS SYSTEMS

Strokes are probably the most common disease associated with hypothermia in the elderly. They may alter the delicate thermostat function of the brain, as well as decrease physical activity.

Figure 7 shows an 80-year-old man admitted to hospital with a temperature of 28°C. He had had a stroke a month before, was living in a very cold room, and was also found to suffer from a rare congenital disease of a part of the brain, the hypothalamus with hypogonadrophic hypogonadism (Kalman's syndrome). The hypothalamus is thought to contain the temperature control centre, or thermostat.

Mental illness, especially dementia, is also often found in hypothermic patients. Demented patients may be unable to respond to a cold environment with the usual behaviour such as turning the heating up and putting on an extra layer of clothes. The thermostat in their brain is also often impaired. Spinal cord and peripheral nerve diseases may also produce hypothermia, by decreasing the patient's mobility and impairing the vasomotor responses.

DRUGS

Especially those acting on the central nervous system. Figure 8 is an example of excessive drug use in old age. The drugs in the basket were brought in with the patient. They may impair the subject's awareness of cold, as well as temperature control. Ethanol in excessive doses not only dulls the senses, but also causes profound vasodilation in the skin. It also interferes with sugar metabolism. It thus causes decreased heat production, increased heat loss and clouding of consciousness. Alcohol abuse seems to be on the increase in this country, even in the elderly population.

The major tranquilliser drugs are very effective in lowering body temperature, and some were indeed used for that purpose in a therapeutic setting, such as heart surgery in recent years. Anti-depressant and minor tranquilliser drugs as well can cause hypothermia. I have seen an 86-year-old woman become hypothermic after a single dose of Nitrazepam 5 mg in a hospital ward during a heat wave when her room temperature was 27°C.[5]

Careful clinical examination of the elderly patient found to be

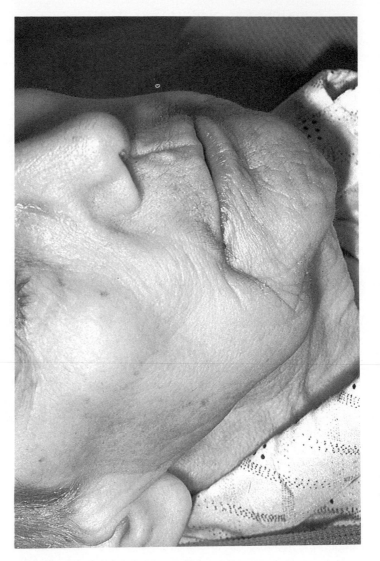

Fig. 7.   80-year-old man admitted to hospital with hypothermia and also found to suffer from hypogonadrophic hypogonadism (Kalman's syndrome).

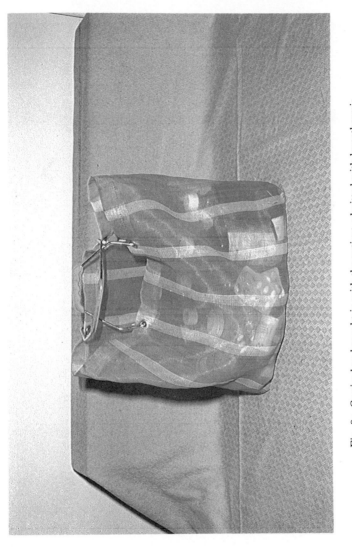

Fig. 8a. Carrier bag brought in with the patient admitted with hypothermia.

M. IMPALLOMENI

Fig. 8b. Drugs contained in carrier bag.

hypothermic often reveals one or more disorders which may have caused, or precipitated, the condition.[6] The commonest conditions associated with hypothermia in the elderly are diseases of the central nervous system, especially dementia and cerebrovascular accidents, in which the thermo-regulatory centre in the hypothalamus seems to be less efficient than it should. Hypothermia occurs most often during winter months. Hypother-mic patients cannot be easily classified according to the first table because they often show features of both primary, secondary and accidental hypothermia in varying degrees.

Severe hypothermia is often a terminal event common to a number of serious diseases in extreme old age. Nevertheless, physicians dealing with the elderly should try to disentangle and diagnose the multiple disorders which may underline this condition. This may be difficult, but rewarding to find; not only the correct management of the present hypothermic episode depends on clearly identifying its causes, but also its prognosis and the prevention of similar episodes in future.

REFERENCES

1. Royal College of Physicians of London (1966) Committee on Accidental Hypothermia. Report. London: Royal College of Physicians.
2. K. J. Collins (1983) *Hypothermia: The Facts*. Oxford: Oxford University Press.
3. M. Impalomeni (1980) Central nervous system disturbance in myxoedema, In: *The Aging Brain: Neurological and Mental Disturbances*, pp. 179–193. Eds G Barbagallo-Sangiorgi and A. N. Exton-Smith. New York and London: Plenum Press.
4. G. L. Mills (1973) Accidental hypothermia in the elderly. *British Journal of Hospital Medicine*, **10**, 691-699.
5. M. Impallomeni and R. Ezzat (1976) Hypothermia associated with Nitrazepam administration. *British Medical Journal*, **1**, 223–224.
6. D. Maclean and D Emslie-Smith (1977) *Accidental Hypothermia*. Oxford: Blackwell Scientific Publications.

Clinical Presentation of Accidental Hypothermia in the Elderly

G. P. J. BEYNON

The Middlesex Hospital, Mortimer Street, London W1

ACCIDENTAL hypothermia can be described as the unintentional lowering of the deep body temperature below 35°C (95°F). It is a serious cause of morbidity and mortality in elderly people, often associated with other underlying disease. Indeed, in most cases multiple factors are involved in the aetiology of accidental hypothermia in the elderly, but exposure to severe cold alone can lead to so-called primary hypothermia in the absence of any predisposing pathological conditions.

The majority of sick hypothermic patients have underlying associated pathological conditions, some of which are listed in Table 1. It is therefore important to remember that in the clinical presentation of accidental hypothermia there are usually mixed signs and symptoms of both the hypothermia itself *and* the underlying disease or diseases.

PHYSICAL SIGNS

The appearance of the patient with puffiness of the facial features, cold vasoconstricted syanotic pale skin, husky voice and slow cerebration may mimic myxoedema and indeed this condition may underlie hypothermia in a very small proportion of cases. Normally warm areas of the skin such as the axilla or groin may be slab-like cold to the touch, the so-called cadaver sign. Generalised oedema may occur and erythroderma and purpura have been noted. The intense vasoconstriction may cause gangrene of the extremeties.

35

TABLE 1. Pathological Conditions Associated with Accidental Hypothermia

Endocrine:	Myxoedema
	Hypopituitarism
	Diabetes mellitus
Neurological:	Hemiplegia
	Parkinsonism
	Wernicke's encephalopathy
Locomotor:	Arthritis
	Immobility following fracture especially of the femoral neck
Mental:	Confusional states
	Dementia
Infections:	Bronchopneumonia
	Septicaemia
Circulatory:	Cardiac infarction
	Pulmonary embolism
Drugs:	Phenothiazines
	Antidepressants
	Hypnotics
	Alcohol
Miscellaneous:	Steatorrhoea
	Exfoliate dermatitis
	Paget's disease

In the cardiovascular system hypotension and bradycardia are the major presenting features, in about 50 per cent of recorded cases. Dysrhythmias, chiefly bradycardia, occur in about one third of patients, but tachycardia may appear. Crepitations often appear in the lung fields, but of course coexistent chest infection is common and crepitations may appear at the lung bases of normal healthy elderly patients.

Signs indicative of pneumonia are common, but a chest x-ray is mandatory in all hypothermic patients. Severely hypothermic patients always inevitably develop respiratory complications in part due to a slow respiratory rate. The bradypnoea may make detection of signs in the chest very difficult. However, tachypnoea may also occur and Cheyne–Stoke's respiration may accompany severe hypothermia.

In the alimentary system the muscular rigidity of hypothermia may mimic peritonitis, particularly as bowel sounds are often diminished.

Abdominal distension may also occur. It must be remembered that pancreatitis is a common accompaniment to hypothermia illness. Indeed, the author mis-diagnosed the first hypothermic patient he ever saw as peritonitis until the summoned surgeon appeared with a low reading rectal thermometer (the only one on the hospital at that time).

In the central nervous system the cardinal symptom is mental confusion progressing to coma as hypothermia worsens. The pupils may constrict or dilate, but a sluggish or absent reflex is commonly noted. Tendon jerks diminish progressively and may ultimately disappear. The last one to survive is said to be the knee jerk, and planter responses may become extensor. Muscle tone is generally rigid. Shivering commonly disappears, but may persist with quite low body temperatures.

If to this constellation of physical signs, of which some or all may be present, are added the signs and symptoms of underlying disease, the fascination of this condition becomes apparent. The cardinal point in making the diagnosis is to think of it, and of course to use a low reading rectal thermometer. This simple process can be life-saving for many elderly patients.

Management at Home

A. T. BRAIN

Hackney Hospital, Homerton High Street, London E9 6BE

THE elderly and the very young have in common a special susceptibility to the adverse effects of exposure to the cold: both groups are vulnerable. Tiny children have no ability to influence their environmental temperature and lose heat readily to the atmosphere as they have a relatively large surface area in proportion to their weight. In healthy adults only exposure to extreme cold is likely to produce a fall in deep body temperature, but as age advances the ability of the body to maintain its internal environment wanes; this impairment of homoeostasis (the name given to the constancy of the internal environment) reduces the margin for error and forms the background to the increased risks of illness and death which are accompaniments of growing old. Vulnerability is therefore a feature of ageing, and hypothermia is one manifestation of this vulnerability.

Hypothermia is defined as a deep body temperature of 35°C or less: it is a disorder which is associated with a high mortality. The deep body temperature is normally maintained by complex mechanisms largely under the control of part of the nervous system which is called autonomic, as its workings are independent of consciousness and are brought into action automatically. The maintenance of the apparently simple balance (Fig. 1) between heat loss and heat production determines deep body temperature. The normal thermoregulatory centre in the brain is very sensitive to temperature fluctuations and sets in motion changes in heat loss in response to a rise or fall in deep body temperature. Sweating and an increase in blood flow to the skin occur when the temperature rises, and

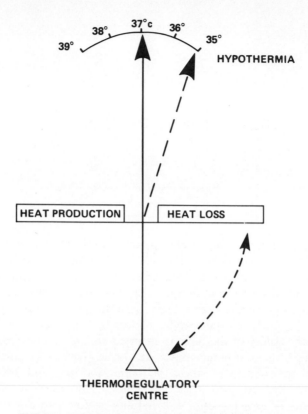

THE BALANCE: THE NORMAL TEMPERATURE IS MAINTAINED
BY A BALANCE OF HEAT PRODUCTION AND HEAT LOSS

Fig. 1.

shivering and a restriction in the skin circulation occur when the temperature falls. Figure 2 illustrates how further stability is added to the unconscious control mechanisms before homoeostasis comes under threat with increasing age. To be young is usually to be healthy, independent and not isolated socially. Healthy adults can respond to a drop in environmental temperature by taking exercise and warm food and by putting on more clothes; the majority can control or choose environmental temperature and are unlikely to be left isolated and cold.

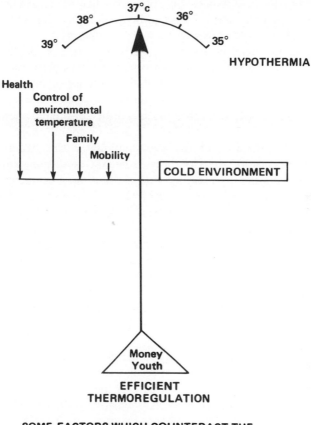

SOME FACTORS WHICH COUNTERACT THE
EFFECTS OF COLD

Fig. 2.

The practice of medicine is concerned with the treatment of people who are ill; it is not about the management of illnesses as if they occurred as separate phenomena unrelated to the people who suffer them. There is, therefore, no such thing as the management of hypothermia at home any more than there is treatment for pain, arthritis or cancer. The understanding of this simple concept is especially important in the medical care of the elderly who, as individuals, can often be helped greatly by modern medicine and surgery and yet tend to be particularly vulnerable to the ill-effects of inappropriate investigation and treatment.

Whenever an elderly person is not well, the nature of the illness should be investigated so that remedies appropriate for that particular individual can be applied. The presentation of illness is often insidious in the elderly because of the background of the ageing process and several diseases often coexist in one person. Hypothermia has to be considered, particularly if the patient is especially at risk, as a cause—or complication—of illness. Given an understanding of the presentations of this condition — an otherwise unexplained deterioration in health, confusion, falls, impairment of consciousness — together with the finding of characteristic signs, particularly the cold feel of the covered parts of the body (the temperature of the extremities is often a poor guide and may be misleading), the diagnosis should not be difficult as long as there is an awareness of the possibility, and the temperature is recorded properly, using a low-reading thermometer. The treatment — the management — follows diagnosis.

First aid should take the form of increasing the environmental temperature, wrapping the patient in blankets (preferably a space blanket), avoiding direct heat in the form of hot water bottles or electric blankets and giving a warm drink if consciousness and the ability to swallow are not impaired.

There are people who develop hypothermia as just one manifestation of a terminal illness; this fact must be recognised and accepted. Distinctions have to be made between forms of care as the quality of life of someone who is dying can be impaired by misguided attempts at cure. If their needs can be met, then it may well be proper for such patients to be looked after in their own homes and management should be directed towards the prevention of suffering, the relief of symptoms and the provision of support for the carers. When a decision is made to give active treatment as opposed to purely symptomatic relief to an elderly person with hypothermia, then arrangements should be made for urgent admission to hospital, the patient being transported with as little disturbance as possible and in a heated ambulance.

The mainstay of care in the home is prevention, and to attempt prevention one must understand the increasing risks to which elderly people are subjected (Fig. 3). When hypothermia occurs it is commonly

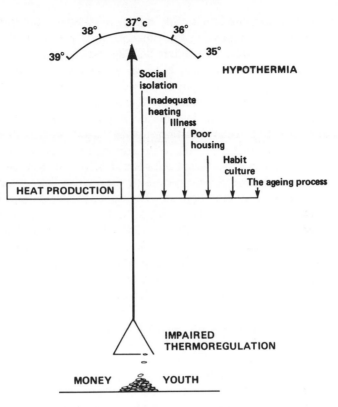

HYPOTHERMIA: THE VULNERABILITY OF ELDERLY PEOPLE

Fig. 3.

because of the cumulative effects of several different factors. The ageing process itself reduces the efficiency of unconscious thermoregulation and many illnesses (and some drugs given as treatment) may impair this efficiency still further so that hypothermia can develop in the absence of a very cold environment. Deep body temperature at 37°C is well above the temperature of the average warm room. Old age and many common illnesses, such as arthritis in its various forms and strokes, which occur in the elderly are associated with reduced freedom of movement and so the ability to take exercise is reduced; similarly, impaired mobility or failing vision may make it impossible for someone to carry or use fuel such as coal or paraffin, light stoves, or regulate heating controls. Many elderly people live in homes without central heating, many live in homes which are poorly insulated and draughty. In addition to these threats to the maintenance of internal warmth there may be others: reduced awareness of cold; malnutrition with loss of the insulation normally provided by subcutaneous fat; social isolation through lack of friends or family; poverty as a contributory factor to inadequate heating. There seems little doubt there is also a significant element of habit and custom among the present generation of elderly people in the British Isles which leads to many living in a cold environment. The notions that it is healthy to sleep in a cold room, that it is extravagant to heat rooms which are unoccupied (even for short periods during the day) seem widespread and are not confined to people whose choice of action is limited by lack of money. Such things may not matter too much when health and youth are present — once these have gone the dangers of a low surrounding temperature grow in importance. Morbidity and mortality increase in the cold even before the point of hypothermia is reached.

An understanding of the many factors which may lead to the development of hypothermia in an elderly person gives clues as to the preventive measures most likely to be successful. It is encouraging that in recent years there has been a much greater public awareness of the vulnerability of elderly people to the ill-effects of cold. Many agencies, both statutory and voluntary, have made significant contributions in terms of education and the provision of practical help, but there is no doubt that large numbers of frail, aged people remain at risk, particularly those who:

1. Live in low environmental temperatures. A survey in the winter of 1971/72 showed 75 per cent of the elderly to be in rooms below 65°F, 54 per cent were in rooms below 60.8°F (16°C). In another survey (of council houses in Somerset) minimum bedroom temperatures were found to be commonly only 2° or 3°C above the outside temperature. People in low environmental temperatures are often in receipt of supplementary benefit, an indication that they are among the oldest as well as being the poorest members of the population; this group of people probably includes the most frail members of society.

2. Have had previous episodes of hypothermia. Survivors commonly show evidence of impaired thermoregulation and other autonomic disturbances.
3. Are ill, including those who are "confused".
4. Are immobile.
5. Are socially isolated and frail.

The groups overlap.

The majority of risk factors require social rather than medical remedies to abolish or mitigate them. As an ideal group of measures the frail, vulnerable elderly should be housed in warm rooms; they should have enough money to allow them to eat well and heat their homes adequately; they should be visited, helped and cared for in the most appropriate manner according to their needs and individual wishes if they are isolated or no longer independent.

Management in Hospital

G. C. J. BENNETT

The Geriatric Unit, The London Hospital (Mile End), Bancroft Road, London E1 4DG

PLANS for the hospital-based management of the elderly hypothermic patient are as varied as the underlying conditions associated with the hypothermia. However, there are themes running through all the different plans which are consistent and are hopefully implemented.

Management of the patient can vary at the point of entry to hospital. Some doctors believe that practically all patients should be managed in an intensive care unit, others that most intervention is unwarranted and even dangerous. Each doctor managing a patient will fit somewhere along that spectrum, bearing in mind that severely ill patients are often unconscious and critical for some hours. Intensive nursing can be, and for the majority is, performed on ordinary wards.

It has been suggested that the management of any hypothermic patient has to take into account the complex metabolic, cardiac and biochemical problems that can occur. Thus the first theme can be viewed as the physiological control needed in management. An airway may need to be established, fluid and electrolytes or acid-base balance may need to be corrected and the heart and vital signs monitored.

Most patients fall into one of four categories during rewarming:

Clinical Course 1   (Favourable)
Rewarming spontaneously and rapidly 0.5°C/hr.
No haemodynamic complications.
Consciousness steadily recovers.
Biochemical and haematological abnormalities correct themselves

47

Clinical Course 2   *(Less favourable)*
    Unexpected hypotension ("rewarming collapse") — (low blood pressure)
    Episodic sinus bradycardia or A–V block — (slow heart rate)
    Profound hypoglycaemia — (low blood sugar)
    Persistent hypoxaemia — (low oxygen level in blood)
    Intercurrent infection

Clinical Course 3   *(Poor prognosis)*
    Normal spontaneous rewarming complicated by —
    Bronchopneumonia
    Acute pancreatitis
    Circulatory arrest

Clinical Course 4   *(Worst prognosis)*
    Core temperature fails to rise or rises only very slowly.
    Signs of "shock" with no cause obvious

The second category probably accounts for the largest number of patients seen clinically.

A few points can be made concerning cardiac arrest (the heart stopping suddenly). At low temperatures the decreased oxygen demands permit longer periods of cardiac arrest with theoretical eventual recovery (up to 2 hours in one reported case). However, electrical defibrillation is usually unsuccessful at core temperatures below 30°C, while some patients may have underlying disease of such severity that cardiac resuscitation would seem inappropriate.

A second theme can thus be established — that of the complications and more importantly their anticipation and avoidance.

Bronchopneumonia or pulmonary oedema (fluid) can be detected by a plain chest x-ray. Similarly abdominal x-rays may show bowel dilatation or free gas, indicating ileus or perforation. Blood sampling (arterial because of venous pooling) may suggest infection, kidney disease, sugar abnormalities, incriminate pancreatitis, hypothyroidism, etc. Assessment of blood gases may be necessary (corrected for temperature). ECG may show characteristic bradycardia and the J-wave.

Many of the complications (fluid shift, hypoxia, hypercalcaemia, metabolic acidosis and hypoglycaemia) may correct themselves during the rewarming process.

The third theme is rewarming and again the choice of method is wide — slow peripheral vs fast peripheral vs central rewarming. The ambient temperature for those admitted should be between 20–30°C. At this temperature most patients rewarm spontaneously at about 0.5°C/hr if insulated against further heat loss (in the old days with a few blankets — now with "space" blankets). The patient should be placed into the blanket, silver side up and swaddled, i.e. head included in the covering up to prevent heat loss. The rectal temperature should be monitored, ideally continuously with an electrical temperature probe.

Maclean and Emslie-Smith have formulated the idea of conservative management of the elderly hypothermic patient:

Conservative management   *(Maclean and Emslie-Smith)*
Nurse in ambient temperature 25–32°C.
Ripple mattress, space blanket.
Cardiac monitor, portable chest X-Ray.
I.V. loading dose prophylactic antibiotics (no more until temperature above 32°C).
Monitor rectal temperature, respiration rate and blood pressure.
Monitor arterial blood gases; give oxygen if necessary.
Monitor blood glucose, electrolytes, acid-base balance.
No routine steroids, infusions, vasoactive drugs or thyroid hormone.
Disturb physically as little as possible.

For most patients I still feel that slow external rewarming is the best method, despite the fact that it means prolonging the physiological abnormality for many hours. Hospital staff are conversant with the method and perhaps more importantly survival figures equate with those of the more complicated rapid rewarming techniques.

Rewarming collapse — seen with all techniques — is severe hypotension (low blood pressure), probably as the result of rewarming too fast for that particular patient. This results in a sudden fall in the peripheral resistance with the cardiac output unable to compensate. Traditional plans have involved cooling the patient again until the blood pressure stabilises and then slowing the rate of rewarming. Vasoactive drugs (inotropes) are being used increasingly to "support" the blood pressure while more active rewarming occurs.

Rapid central rewarming techniques are used in some centres. They should only be used by groups experienced in the technique and probably in intensive care unit surroundings.

Rapid central warming techniques
Warm gastric lavage.
Warm colonic lavage.
Warm intravenous infusion.
Warm mediastinal irrigation.
Partial cardiopulmonary bypass with warm fluid.
Peritoneal dialysis with fluid at 38–43°C.

Thus the themes of underlying physiology, anticipating complications and rewarming techniques should guide our management in this very difficult condition.

Discussion

Question: What can be done to minimise the mistakes in identifying deaths due to hypothermia?

Answer: *Prof. Keatinge:*
The first point I should make is that it is inevitable that there will be a number of misdiagnoses unless a post-mortem is done. But it would have to be misdiagnosis on an absolutely enormous scale to explain something like 40,000 deaths. If one insisted, it would be possible to make it a legal requirement for a post-mortem in every case of death where one is not absolutely certain about the cause of death.

I am quite sure there are situations where misdiagnosis is the other way round. There are cases that tend to get a lot of publicity as hypothermia, but sometimes on closer examination turn out to be carbonmonoxide poisoning. I don't want to go into details, but this has obviously happened on a number of occasions in the past.

Question: From the slides that you showed, it appears that the effects of cold after exposure for 1 hour are not as bad as after 6 hours. Does this imply that old people should not stay out in the cold for a prolonged period of time and that relatively speaking a short period might be OK?

Answer: *Prof. Keatinge*
This was the way we tended to react to that data ourselves, but I should be very cautious about this: This was a very mild

51

exposure to cold, and in more severe exposures to cold I
certainly wouldn't at the present time say that nothing would
happen at the end of that time. Probably a combination of both
duration and intensity is important. All this is being assessed,
but it is really too soon to be very dogmatic about this.

Question: Would it be fair to say that elderly people are more vulnerable
to changes in the blood described as they may already have
arterial disease?

Answer: *Prof. Keatinge*
That is basically our explanation of the situation. What you are
suggesting is quite correct, that the changes in the blood
happen both in young and in elderly people. Perhaps to a
different extent in the elderly than in young people. As young
people have normal arteries, these changes are entirely
harmless to a fit, young person. In an elderly person whose
arteries are badly atheromatous, these changes add a substantial statistical risk of precipitating a complete arterial occlu-
sion.

Question: What advice would you give to people who will be having a
long wait for buses in the cold?

Answer: *Dr. Impallomeni*
Cover yourself well, have a hot drink before you go out, but do
not drink alcohol before going out because that could make
things worse by enhancing hypothermia.

Question: We have recently had a case where someone developed
hypothermia having fallen in the bathroom soon after going to
bed at night. We were told that hypothermia developed
because of loss of liquid through making water. Is this possible
or was there some other underlying cause?

Answer: *Dr. Impallomeni*
I find it very hard to believe, and I share your doubts. In a

series of 40 patients described some years ago, 27 were found on the floor after having spent some hours having fallen and being unable to get up. I would suspect that the patient you are describing would have been one of such patients. Some unfortunate people fall and cannot get up again and they develop much more severe hypothermia. I don't think that loss of water really could have caused hypothermia.

Question: Is there any advantage of giving thyroxine to patients with hypothermia?

Answer: *Prof. Keatinge*
Unless there is clear evidence of hypothyroidism, thyroxine has no role to play in the treatment of hypothermia.

Question: I wanted to ask a specific question about room heating. In olden days, we used to have coal fires, with perhaps better conductivity than the gas fires we have now which will only warm the person in front of it. In addition, in most of the council flats in my area there is half an inch of space between the doors and the floor allowing cold air to come in. Nobody seems to pay any attention to such defects.

Answer: *Dr. Brain*
One has to preach a council of perfection because it hasn't been achieved yet and may never be achieved completely. This does not mean that one shouldn't find out what is needed. I think a great deal should be done with the housing — old houses can be improved by stopping draughts and providing better insulation and so on: I feel one needs to move towards providing an ideal sort of heating that can be controlled by elderly people adequately. Perhaps the best way for that is to provide central heating where possible. It is also very important to make sure that elderly people have enough money to pay for the heating of their choice.

Question: We have heard from different speakers that hypothermia can start with mental confusion, change in skin colour and makes you very slow and sluggish. What comes to my mind is that if you were black, and not diagnosed properly, you may be put into a mental institution. It might sound funny to you, but it does worry the black community when you are talking about diagnosis of white people with such classical signs and symptoms. We feel you should at least address what happens to black people as well.

Answer: **Dr. Benyon**
Cold skin is an important physical sign. Whatever the colour, the skin will be very cold. These patients deserve, expect and should have exactly the same treatment as others, and they normally get it. You feel the temperature in the axilla and the groin. If it is cold and they are confused and drowsy, you should take rectal temperature using a low reading thermo-meter. If they are hypothermic, you get them into hospital as quickly as possible.

Prof. Keatinge
It is very difficult to judge by a person's appearance. I can't diagnose hypothermia by looking at somebody. I can look at them and say possibly they are hypothermic, but it is really not a condition that you diagnose by looking at the person. People who collapse on hills are extremely difficult to diagnose. People do collapse, they lag behind the others, they cool down and they are generally regarded as hypothermic, but if you do measure their temperature, it is surprising how seldom they are hypothermic — they are just exhausted. They may be hypothermic and as a last resort you can only tell by measuring the temperature in à reliable way.

It is a point of some interest that in Sweden, where they have low temperatures outside, it is illegal to certify anybody dead without trying to warm them up. A doctor who diagnoses death without measuring temperature appears in the Courts. The rationale behind this policy is that it is very difficult to assess the signs of life in a severely hypothermic person. Even very young people who have just been exposed to severe cold have slow and feeble heart beats, peripheral pulses are difficult to feel because the arteries are constricted with slow breathing giving the impression that they are dead when they are not. In addition, if people have actually died of hypothermia, you can still revive them to normal health if you start within an hour or so, and in extreme cases up to 2 hours.

Question: Is there any actual data on the admissions of elderly patients with hypothermia looking at whether they live on their own or with other people? Whether, in fact as one would suspect, the incidence is much higher in those living alone?

Answer: *Prof. Keatinge*
Can I ask if anybody has any data on this? I think we have to believe your supposition, and it is very reasonable to suppose that this is the case. I don't know of any data.

Question: Have you any experience with the use of homeopathic medicines in cases of elderly people with hypothermia?

Answer: *Prof. Keatinge*
Homeopathic medicine is given in such small amounts that it doesn't have any side effects. The question is whether it has any other effects. I am sure that most of the medical people here would agree that it is very important not to give people active medicine unless there is a very specific indication, and then to give it for the shortest possible time. If there is no active medicine that will improve somebody and if the patient or relatives are pressing for treatment which is a very common situation, there is certainly a strong case for such things as homeopathic medicine which will at least be guaranteed to do no harm. I produce that as a possible approach.

Question: I am concerned about some of the information I have heard today. Many elderly people attend day-centres or a luncheon club 5–6 days a week. While waiting for buses they are exposed to cold, the effects of which are counteracted by a hot drink and/or meal. The whole process is then reversed on their way back home by waiting for transport. As most such people live alone, their house is usually cold at least for some time after they return. I wonder whether giving a hot meal a day is going to be sufficient or this sort of irregularity is going to undermine their health to a greater extent than would otherwise occur.

Answer: *Prof. Keatinge*
We are very interested in just this sort of problem. What is the actual pattern of cold exposure in most elderly people under different circumstances? I don't think we know the answer.

Question: What is the long term effect of repeated daily exposure to cold during winter?

Answer: *Prof. Keatinge*
I think this is a challenging question and is the sort of thing we have to get the answers to. We are at present collecting information from groups of people with different types of activity and different kinds of house heating. We hope to get some information on such things as car ownership which has been going up considerably over the years. This may be a very important factor in reducing the chances of exposure to cold when you are out of doors. We really don't know the answer to your question at the moment.

Question: People like myself and other pensioners should apply pressure on the appropriate authority to increase the number of buses so that pensioners don't have to wait half an hour for a bus.

Answer: *Prof. Keatinge*
I very much take your point. Of course, there are a lot of things. If we know the information, then it becomes possible to suggest a solution.

Question: The information on the relationship between hypothermia and
              other illnesses has been extremely helpful. May I ask Dr.
              Beynon to enlarge on the connection between hypothermia and
              generalised psoriasis?

Answer: *Dr. Beynon*
              Any disease of the skin which causes scaling and loss of the
              skin surface increases loss of body heat. For this purpose, the
              skin diseases have to be extremely severe and very generalised
              covering a large surface of the body. Not everybody with
              psoriasis which is a very common skin disorder, is liable to
              develop hypothermia. In 20 years of clinical practice I have
              never seen a single case. The problem of covering a complex
              topic like hypothermia is that by including the rare causes one
              can generate a false air of alarm. I want to emphasise again the
              incidence of hypothermic patients being brought to Middlesex
              Hospital in London is dropping. It is because of meetings like
              this causing a general sense of awareness of the problem that
              we are beginning to prevent the occurrence of serious cases of
              hypothermia. But we have got to keep our vigilance.

Question: Having heard that lack of money is a factor in causing
              hypothermia. With the rapid increase in unemployment, are
              we about to see an increase in this problem amongst other
              social groups particularly children?

Answer: *Prof. Keatinge*
              I doubt if any of the speakers would want to comment on what
              is likely to happen. I think we can only comment on what we
              know has happened: Most of the evidence stated by Dr.
              Beynon shows a tendency for a drop in the reported cases of
              hypothermia. I am sure we all have to be careful about these
              statistics because the diagnosis of hypothermia is not all that
              precise.

              Dr. Beynon
              We can over-emphasise the lack of money, fear of unemploy-
              ment and people being poorer than before. It is largely the

question of awareness of a problem. A well-organised group
general practice with full ancillary support of district nurse,
practice nurse and health visitor can have an age/sex register,
identify the elderly and those at risk from certain obvious
factors for hypothermia. I don't think that money is a great
element here. I think supervision, identification and common
sense measures can prevent hypothermia.

Question: I would like to ask a question on drugs. Recently one day I
found my sister who suffers from Parkinson's disease, confused and falling all over the place. She was taking six different
types of tablets which had been repeatedly prescribed. Her
doctor told me that it was not her responsibility to remove the
unused drugs. Can I ask who is responsible for supervising the
unused pills left with confused elderly patients living alone?

Answer: *Prof. Keatinge*
Does anybody have specific expertise on this question of how
to handle the problem of confused elderly people, as to
whether they take repeat drugs?

Dr. Beynon
The question of repeat prescriptions is very difficult in General
Practice. The average GP's list is about 2500 patients. There is
no doubt that in the 50s, 60s and 70s with the NHS developing
there was a mood in the country that drugs could cure
everything. In the past 10 years we have realised the hazards of
drugs in pregnancy, old-age and children. The attitude of the
medical profession has changed and particularly in my
speciality of the elderly the less prescribed the better. There
are various systems for monitoring repeat prescriptions in
general practice. Unfortunately, the workload of the GPs is so
vast that occasionally these repeat prescriptions slip through
the net. It is a difficult problem. As to the responsibility of
removing drugs from a patient's home who lives alone, there is
no clearcut answer to that. It is clearly not the GP's
responsibility to tour homes and recoup drugs. This is

something which really ought to be properly addressed throughout the country. It is in a sense the family's responsibility, but the family members are sometimes far away and don't visit. Some patients do not have families. I do not know what the answer to that question is.

Question: We in the housing research unit have been correlating the figures for hospital admissions for certain illnesses against temperature fluctuations. So far we have not been able to obtain statistics on hypothermia, I wonder whether any of the speakers might like to comment on the relationship between hypothermia and other illnesses. It is difficult for us to disentangle the difference between the effect cold has on certain illnesses for which patients are admitted to hospital, the role of cold precipitating the illness and admissions to hospital, and whether hypothermia should be considered as a distinct illness. We find that bronchitis and other diseases are recorded when people are admitted to hospital, but apparently not hypothermia.

Answer: *Prof. Keatinge*
I think the point really is that we have very much better statistics on mortalities than we do on illnesses. There is no systematic method of assessing patients who may notify their doctor that they are ill. Even after going into hospital, it is not easy to collect statistics on admission to hospital nationwide for a number of disease patterns. We do have very much better evidence on mortalities and that is basically what was shown earlier on this morning, i.e. the effect of air temperature on mortality from various conditions recorded as the cause of death.

Dr. Collins
Could I respond briefly to the last question? Basically, there are two important temperatures that elderly people are generally concerned about. The first one is the temperature for comfort and second is a lower temperature for health, in

whatever way one defines health. There is a certain amount of evidence now from the Building Research Establishment and from various research institutions throughout the country that a temperature between 18°C and 23°C for a normally clothed person is satisfactory as far as comfort is concerned. If you drop the temperature below 18°C, the next probable temperature to consider is round about 16°C. According to the Factories Act, this is the lower limit for health in people working in factories and it is probably true for home environment as well. Below a temperature of 16°C, there is a trend for an increase in respiratory illnesses. The next important temperature point is round about 12°C. I think 12°C is the sort of threshold limit at which we start to notice a rise in blood pressure. Most reasonably healthy people can withstand temperatures down to 12°C without thermoregulatory trouble. It is not until one gets to very low temperature, perhaps 5°C or even lower, when the thermoregulatory system is unable to cope with the cold stress. That basically is a rough guide to the temperatures that one should aim at in the home for health and comfort.

Question: I work at a retirement village for licencees. There are 280 of them and they all die on the premises. Normally about 30 or 35 die in the course of the year. Some of them are brought in ill from outside. So far this year we have had about 50 deaths. I wondered whether this was due to a carry-on effect from the last year's cold winter. Whether the health of elderly people who have become cold and survived continues to deteriorate?

Answer: *Prof. Keatinge*
I think this is a very interesting comment. I think it is one of the many things on which we need to collect information. I don't know of any answer to that question, but it's one of the things which should be coming up from the surveys under way.

Section II

HYPOTHERMIA

SOCIAL ASPECTS

Why do the Old go Cold, and What Can We Do About It?

Scottish Council for Voluntary Organisations, 18/19 Claremont Crescent, Edinburgh EH7 4QD

IT is encouraging to know that the deaths from certified hypothermia in England and Wales have approximately halved since 1979. This does not, however, mean that hypothermia as a threat to the elderly and the very young has been contained or defeated. A great deal of work needs to be done to reduce the risks of cold in the elderly.

In order to define the magnitude of the problem facing us, let us examine the question "Are the old cold?" The first major national survey of heating and the elderly was carried out by Wicks in 1972 and published in 1978. As it happened, the winter in 1972 was rather mild. He studied a sample of 1000 pensioners and found that:

90 per cent had morning living room temperature below the old DHSS standard of 70°F.

75 per cent had morning living room temperatures below the Parker-Morris standard of 65°F.

54 per cent had morning living room temperatures below the legal minimum set out in the Offices, Shops and Railway Premises Act (1963) of 60.8°F.

Furthermore, he found that in 84 per cent of cases, morning bedroom temperatures were below 60.8°F. Perhaps the most worrying finding of this survey was that 9.6 per cent of pensioners were "at risk" from hypothermia. That is to say, they had a deep body temperature of 35.5°C, within 0.5°C of the clinical definition of hypothermia. This will put 800,000 people at risk in the United Kingdom.[1]*

A similar survey carried out by Primrose and Smith[2] in the Govan area of Glasgow found that out of a sample of 220 pensioners:

> 97 per cent were below the DHSS standard.
> 84 per cent were below the Parker-Morris standard.
> 61 per cent were below the Offices, Shops and Railway Premises Act's minimum recommentation.

Primrose and Smith summed up their findings by saying: "Although few of the elderly are hypothermic many continue to live at temperature levels well below their comfort zone." The finding that around 50 per cent of the pensioners are likely to heat only one room in the house during the winter[3] may help to explain the findings of above studies.[1,2]

Clearly there is a need for further research here. In particular we know little of the connection between cold weather and *cold-related illnesses*, such as bronchitis and certain heart diseases. Preliminary work done by Markus and Wooley[4] at Strathclyde University suggests an increase in admissions to hospital for such illnesses 2 or 3 days after a cold spell.

Prevention of cold-related illnesses amongst the elderly by encouraging retirement into the warmer climates is an attractive theoretical proposition. It is true that United Kingdom climate is not very friendly in winter, but we should note that other countries with climates comparable to or worse than our own such as Canada, Sweden and Denmark do not have a 20 per cent increase in death rate amongst the elderly during winter. Therefore, climate does not *explain* our poorer performance here.

Let us now consider some of the socio-economic factors likely to contribute to our poor performance in preventing hypothermia and other cold-related illnesses among the elderly.

First of all, the elderly are, by and large, poor. Around 1 in 4 are at or below Supplementary Benefit level and 60 per cent of pensioners are in the

---

*Superscript numbers refer to References at end of chapter

bottom quarter of income groups.[5] We can see how badly off some elderly are when we realise that 80 per cent of pensioners on Supplementary Benefit get help with their heating through additional payments. The number of such payments has risen from 230,000 in 1972 to more than 2 million in 1982 and the majority of such payments go to pensioners. Within the elderly population there is a considerable variation in income levels. Work carried out by Age Concern England a few years ago suggested that average income falls from £102 per week at 60–64 to £29 per week for the over 75s. Because they are poor the elderly spend a larger proportion of their income on fuel. On average they spend 10.5 per cent as compared to a national average of 6 per cent. It is the third major item of expenditure after food and housing. Work carried out by Isherwood and Hancock at the DHSS indicates that some elderly people spend 18 to 24 per cent of their income on fuel. Despite the fact that older people spend a greater *proportion* of their income on fuel, they still spend *less in absolute terms*. Put simply, 10 per cent of £30 is still less than the 3 per cent of £250 that a rich person might spend.

Other poorer groups are in the same position as the elderly with one interesting difference, which is that pensioners do not on the whole like to run up debts and get disconnected. They tend to stop using fuel rather than go into debt.

One of the hard facts of the last 10 years is the sharp increase in fuel prices. The Department of Energy expect fuel costs to double in real terms by the end of the century. It is quite clear that all fuels, except gas, have risen in price faster than inflation over the last decade. The era of cheap fuel is over. One consequence of this is that the elderly are now able to buy less fuel than they were in 1973. Indeed, we know that pensioners spent the same proportion of their income then as now, so they must also be buying less fuel as the example below shows.

	1974	1982
Income	£10.00	£38.50
Per cent spent on fuel	10.5% (£1.05)	10.5% (£3.87)
Fuel cost per unit	1p	4.2p
Units	105	92

Apart from the fuel costs, British housing conditions are poor compared with other European countries and North America. Insulation standards are twice as high in some of these countries. We still have 30 per cent of British housing without loft insulation and a mere 12 per cent are up to the government's recommended standard. This means an enormous waste of money. It is estimated that up to 75 per cent of heat and thus expenditure may be wasted in a poorly insulated house. Many elderly people are living in the worst housing conditions without the benefits of central heating. A recent report presented to Kenneth Baker argued for an extra £20 billion to bring local authority housing up to standard.

Part of the problem here is the way in which we have been the victims of the prevalent view in the 1950s and 60s that energy would be cheap and getting cheaper. This in turn led to houses being designed for cheap energy rather than being energy efficient, and we now find ourselves locked into the situation where energy is expensive and yet house design does not reflect that expense.

None of this speaks well of our long-term energy strategy. We find ourselves in the position of increasing energy supply through building expensive new power stations rather than looking at the conservation alternatives. Interestingly enough in the United States the power utilities board has decided that it is cheaper to invest in energy saving for their customers than building new power stations. In Britain we have the odd position of some £1.4 billion being spent helping people with their heating bills through the DHSS, but only £18 million spent on long-term conservation work which would help cut these bills down.

The current economic policy behind fuel pricing is to push up prices, especially gas, for reasons that have little to do with the needs of the industries themselves. The consumer councils constantly make this point. One cannot help feeling that the current fuel price increases are being used as a way of raising additional revenue for the government — a sort of back door tax. As a tax it is thoroughly regressive and hits the poorest hardest of all.

The final area I want to look at is that of demographic change. Quite simply there will be more old people around in the future as we succeed in increasing life expectancy. In 1901 there were 2.4 million people over pensionable age out of a population of 38 million. In contrast, there were 12 million people under 16. By 1981 the number of under 16s had remained the same, but the number of pensioners had quadrupled. Women pensioners now outnumber girls under 16. The proportion of female pensions has increased from 6 per cent in 1901 to 16 per cent in 1981 and is expected to increase to around 18 per cent by 1991. We must not forget that the word "pensioner" hides a wide range from 60 to 90 and beyond. The numbers of older pensioners, those over 75 and those over 85, are due to increase rapidly in the next 20 years. They are the people who are most at risk from cold and hypothermia. In 1981 one person in 104 was over 85 and by 2001 it is likely to be one person in 65.

It is far easier to see what the problem is than to come up with any magical solution. Many aspects of a sensible solution lie outside our control in the hands of the government. Perhaps I can look at three broad areas for improving matters.

First of all there is the question of *money*. We need to make better use of the money that is currently available: How often are we reminded that £1 billion in benefits remains unclaimed each year and that 600,000 pensioners do not claim the benefits they are entitled to. Many local authorities have supported take-up campaigns with a high degree of success, but more still needs to be done.

Even if maximum use was made of the money that is available, we would still need to recognise that the current rate of pension and benefit is not enough. The national element for heating in Supplementary Benefit is not adequate for winter fuel expenditure. Raising the value of the pension is highly desirable for many reasons, though we should be careful in not jumping to the conclusion that if done an increase in pensions will automatically solve the problem of cold conditions. One problem that should be recognised is the climate differences within the United Kingdom. It costs 20 per cent more to heat a house in Glasgow to the same standard as an identical house in Bristol and 30 per cent more in Aberdeen.[7] Last year we had the bizarre occurrence of special cold weather allowances being paid out in the South of England but not in Scotland where the weather was actually colder there! The reasoning behind this is hard to understand. If the weather is wetter, windier and colder in Scotland, then that should be reflected in the level of payment for heating. The National Right to Fuel Campaign has done some interesting work on constructing a special "cost of warmth" index which might be worth further investigation.[8]

There is, however, only limited use in giving people more money if they still live in badly insulated homes and do not know how to get the best out of their heating system. We need projects that will give better advice to people and provide it in a way that people can understand and relate to it. It is surprising that over 80 per cent of private expenditure on energy conservation is for double glazing — the least cost-effective measure available. It demonstrates the power of advertising which should be directed at cheaper and cost-effective methods for energy conservation. One feature we can certainly adapt from the United States is the Energy Utilities Board (British Gas or the Electricity Board in the United Kingdom) having to give more energy audits to consumers.

Let us now look at the value of local authority insulation projects. In 1981 there were a few pioneering projects. Thanks largely to the support from the Department of Energy and local authorities, there are now 172 such projects in the United Kingdom (35 in Scotland). One of these projects, Heatwise Glasgow, is a model example for co-operation and collaboration between the voluntary sector and a local authority which is beneficial to all concerned. Heatwise has twelve local projects in Glasgow, each run by a local committee. It employs over 300 people and has a turnover in excess of £2.5 million. Glasgow District Council provides £500,000 and the rest comes from the Manpower Services Commission, the European Social Fund and the Department of Energy. The benefit for the local authority is that in every £1 it invests, it brings in £4 from elsewhere.

A more serious threat is the current review of the social security system presently under consideration. The proposals to abolish single payments and weekly additions (which will include payments in draught-proofing and heating) will have a devastating effect. The social fund which is intended to mop up such cases cannot possibly cope with the volume of need. I feel that unless real changes are forced on the government, the heating needs of the elderly will not be met. It is one of those oddities which makes life so challenging that these harmful changes will be announced during the government's own Energy Efficiency year.

REFERENCES

1. M. Wicks. *Old and Cold Hypothermia and Social Policy*, pp. 7ff. Heinmann, 1978.
2. W. Primrose and L. Smith. Oral and environmental temperatures in a Scottish urban geriatric population, *Journal of Clinical and Experimental Gerontology*, **4**(2), 151–165 (1982).
3. P. Townsend, *Poverty in the UK*, p. 286. Penguin, 1979.
4. T. Markus and T. Wooley. Unpublished observations, 1985.
5. A. Grimes. Poverty and Scottish pensioners. In R. Cook and G. Brown (eds.), *Scotland: The Real Divide*, pp. 136–144. Mainstream, 1983.
6. B. C. Isherwood and R. M. Hancock, *Household Expenditure on Fuel: Distributional Aspects*. Unpublished observations, 1979.
7. T. Markus. Fuel poverty in Scottish homes. *Architects Journal*, pp. 1077–1082 (1979).
8. B. Boardman, *Cost of Warmth Index*, pp. 3ff. National Right & Fuel Campaign, 1984.

The Preventive Role of Age Concern

M. COLLYER

Age Concern Greater London, 54 Knatchbull Road, London SE5 9QY

DEATH from hypothermia is a bitter commentary on a civilised society. A much greater problem, however, lies in the fact that large numbers of elderly people are immobilised and made miserable by being cold. In these circumstances hypothermia might well be seen as a blessed relief. Although provision of information and support have a valuable preventive role, the fact remains that many elderly people face a stark choice between burning fuel which they cannot afford or struggling to survive in freezing conditions.

Half of the nearly 10 million pensioners in the United Kingdom live on or below the poverty line. Approximately 3 million live alone; and well over 1 million are aged 75 or over. Nearly three-quarters of their expenditure goes on housing, fuel and food. Elderly people spend nearly twice as much of their income on fuel as the rest of the population, due to the fact that they spend more time at home and need heat for most of the day. Obsolete heating equipment is a problem, as it can be both expensive and inefficient, and nearly half of pensioner households have no central heating. It is no wonder that on average elderly people spend less on heating their homes in winter than most families do in the summer. Nearly 40 per cent of elderly people do not heat their bedroom at all in the winter.

The difficulties which older people face in trying to keep warm often preclude them from participating in the mainstream of life. The prevention of hypothermia depends on promoting good health and activity in old age. Age Concern responds to these issues in a number of ways.

THE AGE CONCERN MOVEMENT

The Age Concern movement is a network of independent and self-governing local groups, supported in London by Age Concern Greater London and on matters of national significance by Age Concern England. There are about 1000 autonomous groups in England. Each of them forms its own policies within a generally agreed national framework. Some groups, particularly those operating over a limited part of a borough or health district, still use the name "Old People's Welfare Association", but all of them are a part of the Age Concern movement. The main objectives of the Age Concern movement are to provide effective direct services to elderly people, including advice and information for older people and their carers; to pioneer new projects and research; to promote partnership in joint social planning — a co-operative approach which involves promoting links across departmental and organisational boundaries; and social advocacy, which includes both public education and campaign work. Any discussion of the heating problems of elderly people is informed by each of these four objectives.

YEAR-ROUND SERVICES TO ELDERLY PEOPLE

Direct services provided throughout the year perform a preventive role. Support provided in the home through voluntary visiting — often undertaken by retired people — or through relief for carers will pick up danger signals such as poor diet, insufficient heating appliances, or the under-used. Volunteers who are experienced swimmers are particularly day centres and clubs where a hot nourishing meal is provided; and informal "Pop-Ins" where the more active pensioners can buy themselves a snack and a hot drink. These centres provide warmth, companionship and nutritional advice, together with various services which help pensioners to keep active. For example, some health authorities arrange for a qualified chiropodist to visit Age Concern day centres regularly to look after the feet of the elderly people who use the centres. Many day centres sell welfare foods at low cost to pensioners, many of whom find difficulty in getting to the shops. All these year-round services encourage self-help which is crucial to good health in old age.

DIET AND EXERCISE

In order to keep warm people need to eat well and keep active. However, if you live on a reduced income it can be difficult to choose a sensible diet. Age Concern groups try to overcome these difficulties in various ways. They may offer nutrition leaflets and recipe booklets, sell cost-price nutritional food, and food packaged in small quantities. A number of groups run "cook and eat" classes, particularly useful for widowed men who have to start catering for themselves late in life. Age Concern Feltham, for example, has a men's cooking club which originated in their weekly social club; Age Concern Hammersmith and Fulham have recently obtained funding for a nutrition bus to visit housebound elderly people; and Age Concern Newcastle has arranged for a dietitian to visit their centre every week to help pensioners plan nutritious meals.

Age Concern helps to promote good health in old age by stimulating activity and campaigning for better health care. There is a great variety of activities on offer, including exercise and keep-fit classes which help to improve circulation and keep people of all ages mentally and physically alert. Age Concern Greenwich runs relaxation classes for Asian women, and pensioners' self-defence groups; Age Concern Southwark organises physiotherapy and skittles for older people recovering from strokes; and Age Concern Durham is famous for its tea-dances and discotheques. Many groups are able to come to an arrangement with a local swimming pool whereby their elderly members are able to attend regular private swimming sessions at times (such as lunchtime) when the pool is otherwise under used. Volunteers who are experienced swimmers are particularly important for this type of work. One Age Concern group recently taught a 90-year-old woman to swim for the first time in her life.

INFORMATION AND ADVICE

Age Concern offers information and advice about welfare rights, benefit levels and ways to make the most of income. Age Concern England publishes the popular *Your Rights in Retirement* which is annually updated at the time of changes in pension and benefit levels; and has also produced booklets on *Your Taxes and Savings in Retirement*; *Heating Help in Retirement*; and *Warmth in Winter*. A number of local groups produce their own booklets on local services which include sections on heating allowances and advice. More specifically, groups organise pensioners' conferences on related issues. For example, Age Concern Southwark recently ran a day conference, *Keeping out the Cold*, organised jointly with Camberwell Health Authority. The programme included an explanation about hypothermia; information on budgeting and benefits; practical ideas for keeping warm; information about heat conservation, sensible eating and exercise, and what Age Concern can do to help. An information pack was distributed to all who attended the conference. It included advice on how to recognise hypothermia and what to do; on draught-proofing; food; Age Concern's services; the new pension and benefit rates; and some useful addresses and telephone numbers.

Pensioners' health days, increasingly popular, also provide the forum for information of this sort. They bring together a variety of professionals and voluntary groups who can each offer expertise and support. They usually provide a nutritious wholefood lunch at low cost together with recipes and culinary advice.

COLD WINTER CRISIS

In periods of winter crisis many Age Concern groups keep a stock of heating appliances and blankets to lend in cases of emergency. Research undertaken by Age Concern Greater London shows that during the winter of 1984/85 most groups distributed their entire stocks of these items. They received extra calls for home visiting as many pensioners felt trapped in their homes owing to icy and dangerous pavements. Requests for home visits and de-icing of pavements contrasted with the popular view in the media that the answer lay in opening up communal facilities.

Age Concern Brent's advice team for housebound people reported lack of food supplies, the need for extra support on discharge from hospital, and frozen and burst pipes as the most serious problems. Help given by this team included claiming payments for essential household repairs, insulation, and the replacement of obsolete heaters; and claiming additional allowances towards special diets, extra heating costs and laundry. The team reported that "If these (allowances) are made harder to get, more people will face extreme poverty; if they are totally abolished it is difficult to see how some people will survive."

INSULATION AND DRAUGHTPROOFING

Sixty per cent of pensioners have no loft insulation and 70 per cent have no draught-excluders. The role of Age Concern is two-fold: to urge pensioners to claim for grants and to refer them to a source of appropriate help. A few Age Concern groups, including Age Concern Salford, Avon and Tyneside, run draught-proofing services themselves. Age Concern Westminster employs a Heating Officer who works with a full-time draughtproofer and a small group of carefully selected volunteers. An important preventive service is provided through regular liaison between the heating officer and the Age Concern welfare workers regarding other problems identified by the draughtproofer. Apart from Age Concern Westminster, older people in London are generally referred to projects in membership of the London Energy Project Forum, funded by the MSC, the DoE and local authorities. These projects offer free labour for draughtproofing. The material costs are often met by a DHSS single payment for pensioners in receipt of supplementary pension who have savings of less than £500. Energy groups also offer loft insulation, hot water tank lagging, and general advice on heating.

At present, elderly people qualify for 90 per cent of the costs of home insulation up to a total of £95. On the face of it this seems generous, but some people cannot afford even 10 per cent of the total bill. In addition, local authority money available under the Homes Insulation Scheme has been cut back from £35 million in 1984/85 to £31 million in 1985/86. A further problem arises through lack of awareness which results in limited take-up from those people who are most in need. Age Concern is campaigning for the maximum homes insulation grant to be increased to £120, with 100 per cent grants to elderly people; and for grants to cover the costs of general insulation and draughtproofing.

DIRECT FINANCIAL HELP

Some Age Concern groups are able to provide modest financial help to cover heating bills at times of crisis. Heating funds have often grown from individual donations, usually sent in place of flowers at a funeral. Age Concern Greater London's own Heating Fund is at low ebb as great demands were made by the winter of 1984/85. Between December and May over £1500 was distributed. The continuation of this scheme depends on further legacies and support.

WARMTH IN WINTER

Age Concern continues to campaign on a range of heating issues. In 1983 Age Concern England mounted a Warm up Winter Campaign which had two policy objectives: the development of a national energy policy; and an increase in the value of the retirement pension. In the short term, Age Concern argues for heating additions to be made available to everyone who is entitled to housing benefit, to have the recent cuts in heating additions restored; and for the total end to the disconnection of pensioners' fuel supply.

Age Concern England has recently published *Left Out in the Cold: The Heating Cost Implications of the Reform of Social Security for Elderly People*. This report poses serious questions about the Social Services Review and forms part of the Age Concern response. In the long term Age Concern believes that an adequate level of income is the right of every elderly person so that he or she can play a full part in the social and economic life of the community, exercising real choice about how money is spent.

In the United Kingdom there are almost 10 million retired people, and within the next five years a further 6 million will retire. These voters have the potential to become an effective force able to influence political decisions about both the 'evel of income and the standard of care available to them. Both in its own work and together with other organisations Age Concern seeks to help to develop this influence.

Statutory Social Services for the Elderly at Risk

N. F. COSIN

Camden Social Services, 156 West End Lane, London NW6

THIS winter a large number of our old people, and tragically a smaller number of our very young, will die of hypothermia. Hypothermia is an indicator in its extreme form of the impact of cold conditions on health. Whilst the number of deaths where hypothermia is given as the certified cause is not very high, and there is considerable official caution over identifying cold as a major cause of death among the old, large numbers of our old people live in very cold conditions.

What can be done for individuals at risk in this way? This article will review the available social services to the elderly in the London Borough of Camden with a brief mention of the wider national issues.

PERSONAL SOCIAL SERVICES IN CAMDEN

In Camden, we have 28,000 residents over 65, of whom 12,000 are over 75. There are over 6000 registered disabled in the borough, with different levels of permanent and substantial handicap. Unemployment runs at 16 per cent compared with a national average of 13 per cent.

The available help to the needy comes in three forms — statutory help from central and local government; voluntary help from charities and voluntary organisations; and a third "invisible" sector who provide by far the greatest amount of help in hours, in effort and in sacrifice, the informal carers. This last category comprises relatives, neighbours and friends, without whose care and work all organised assistance would be completely overwhelmed. Seventy-five per cent of these carers are women.

Camden services have tried over the years to offer resources which will significantly assist people whose lives are constrained by handicap, frailty and material disadvantage. Each service is designed not only to offer direct help to the person in need, but to offer support and relief to that third sector, the informal carers.

Camden's two most comprehensive services for the elderly at risk are the Home Help Service and the Meals Service. 400 home helps visit over 5000 homes (elderly, handicapped or younger families) each week. Far from the outdated image of "cleaners on the rates", they represent the only significant social contact for many of the vulnerable elderly. Doctor, health visitor and the social worker are all vital parts of the team, but it is often the home help whom the confused, elderly person knows best. There are too many incidents of care arrangements in the community actually breaking down when a familiar home help is changed, for us to underestimate their crucial role.

The Meals Service delivers 1500 meals each day and 500 on week-end days, providing further contact and monitoring for frail people on their own. Occupational therapists, whose training is geared to maximising an individual's independence, both physical and psychological, offer a more specialist and harder to get service. High referral rates and long waiting lists make this service one of the most stretched in the borough and means people with serious disabilities living at home can have hospital admissions or lead grossly restricted lives for want of earlier attention. This is a service which has been caught out by hospital policies of early discharge, demographic changes and rising expectations.

Each fieldwork office has a volunteer organiser to co-ordinate the activities of up to 100 people who can offer a wide variety of help to isolated individuals. Where care outside the home for an old person is the preferred option, there are Day Centres or Homes for the Elderly, which are accommodating an increasingly frail and mentally infirm population.

In 1984 welfare rights officers were appointed to raise the take-up benefits to which the vulnerable people are entitled.

Three schemes pioneered in Camden and not yet universally available are worth mentioning here. These are:
1.   Adult Care Scheme.
2.   Peripatetic Care Scheme.
3.   Special Home Care Scheme.

The Adult Care Scheme involves the recruitment of individuals or families who are prepared to take a frail person into their home on a long-term basis. It is sometimes thought of as a kind of adult fostering. They receive a weekly fee for offering accommodation and care to a person who may be mentally infirm, disabled or elderly frail. The important advantage is that it prevents the person having to enter an institution and provides a more individual style of care for that person. Since its inception in 1974, it has grown to accommodate over 150 individuals in 82 private households.

The Peripatetic Care Scheme employs officers of the department who can go into a home daily or live in and provide care for an individual or for children where the normal carer is unable to continue. Examples include a devoted spouse who is near exhaustion point taking a week's holiday while a peripatetic care officer lives in the home and looks after the demented partner. Another example is of an officer living in a family home of a single parent, caring for two children while mother is confined with a third pregnancy.

A more recently developed scheme is a Special Home Care Scheme. This is designed to provide severely disabled people with care helpers in their own homes. It differs from the other schemes in the following ways:

(A) The care helpers are recruited for a particular client who participates in the recruitment.
(B) The provision is as long term as the client requires it, whilst other such domicilary schemes (peripatetic care or the voluntary Cross Roads Scheme) are essentially respite or emergency measures.
(C) Each arrangement will be for an individual package of care in which the client is as much in control of the service as the council officer who co-ordinates the scheme. There are 30 such clients receiving this service. Some of these matches arose as informal caring arrangements which have now been supported financially by the scheme.

Emergency help from the fieldwork teams is available for old people without heating or cooking facilities from both Social Services and Camden Old People's Welfare Association, in the form of a loan of heaters, blankets, quilts, small cooking rings.

Where does the local authority social worker fit into these schemes? Our

Help the Aged Spreads the Word

P. L. BARON

Help the Aged, St. James's Walk, London EC1R 0BE

WITHIN the United Kingdom our national programme is to concentrate on four main areas — day centres, minibuses, sheltered housing and Lifeline our emergency communication system. Our support is channelled through churches and other community-based organisations, but a substantial amount is given to local Age Concern groups who have a programme and network of community care second to none.

Hypothermia is as much a social problem as a clinical one, with those who are most at risk living in bad housing conditions. The Duke of Edinburgh's inquiry into British housing showed that 43 per cent of all the "unfit" properties in England were occupied by elderly people. Admittedly for some the situation has changed, but only recently, for as short a time ago as 1972, 3 out of 4 elderly people were unaware of the extent of supplementary benefits and only 11 per cent at that time were actually receiving extra heating allowances. Today estimates show 90 per cent of those entitled are now receiving this extra allowance. But nearly 80 per cent of old age pensioners fail to qualify for supplementary benefit, often by a narrow margin and may have to struggle without any heating allowance.

It is a fact: that old people spend more of their income on fuel than other age groups

It is a fact: that on average single pensioner households spend 13 per cent

of their income on fuel costs, whereas the average UK household for all
ages, spends 6 per cent of its income on fuel.

It is a fact: that only one third of old people are benefiting from the
efficiency of central heating, whereas the national average for all
households is 50 per cent.

With this knowledge therefore it is the policy of Help the Aged to press for
an adequate heating allowance for retired people dependent on supplementary pension. We were pleased with the development of the heating
addition, and recognition of the needs of the very old, but last Autumn our
delight quickly turned to horror when the government took the money
back again by another route leaving 1½ million pensioners worse off!

Help the Aged is lucky enough to have built up over the years a
veritable army of supporters in school children of all ages. As part of our
schools programme our community staff always include in their talks
something about the causes and effects of hypothermia. The older
children, when they go home-visiting, are instructed to look for tell-tale
signs of room temperature, lack of food in the cupboards, heating
appliances not switched on, apathy and sleepiness, difficulty in breathing.
If any of these signs are present, they contact their organising teacher
without delay.

At the heart of the educational work of Help the Aged is the premise
that the quality of life in old age in any society depends on the attitude of
younger members towards old age. Negative attitudes will colour the
social policies affecting the lives of older citizens. Psychological and
medical research has shown that severe mental and physical decline is not
an inevitable consequence of ageing. We have increased overall life
expectancy to 65 + 12 years for men and 60 + 20 years or more for
women. Yet the general level of expectation of what life can hold for an
older person is very low. It has somehow become accepted that older
people need less money, live in older and substandard accommodation, do
not need stimulating occupation and a social life. In this way society has
created "the problems of old age" and created a "class" of old age
pensioners seen as different from the rest of the population. Unless such
attitudes are changed, very little is going to change for our pensioners:
Sixty years ago Lloyd George observed, "How we treat our old people is a
crucial test of our national quality. A nation that lacks gratitude to those

who have honestly worked for her in the past, whilst they had strength to do so, does not deserve a future, for she has lost her sense of justice and her instinct of mercy."

To bring about any change in these attitudes, Help the Aged's Education Department considers it essential to bridge generation gaps and to foster positive attitudes towards old age amongst old and young alike. The loneliness, isolation and the growing poverty of many elderly people means that the need for information and understanding about old age is now greater than ever before.

As a result of our wide range of teaching materials, learning about old age is fast becoming a normal part of the school curriculum. We hope this will increase the number of elderly people looked after by their younger relatives in their own homes. To help the young ones, Help the Aged has recently published a new home economics teaching pack *Nutrition for Life*. It gives basic principles of healthy eating relevant to the lifestyles of elderly in a variety of situations and shows how diet affects our health throughout life.

The Education Department works closely with those involved in planning and training of health and social services employees caring for the elderly. The latest handbook related to this area is *In Our Care*, designed for lay and professional staff. *The Time of Your Life*, Help the Aged's widely read retirement handbook is now in its fifth edition. It is interesting that many older people themselves share society's negative attitudes towards old age. By providing information and suggestions for a creative and fulfilled retirement, we hope to counteract those preconceptions which contribute towards isolation and loneliness in later life. Michelangelo was a remarkable example of a person unrestricted by old age. You will recall he began to work on the design of St Peters at the age of 71!

Provision of day centres, to which Help the Aged is pledged, plays an important part in combatting hypothermia — somewhere to go where it is warm during the day, a place to have a hot meal, company and stimulation. I can still vividly remember a day centre warden telling me some years ago, "It's the devil of a job to get everyone out by 4 o'clock when we close, they don't want to go home. You cannot blame them because they often have to go to an empty house which is cold, and they won't see or speak to another person for hours, possibly days."

We know how a simple fall can put an elderly person at unexpected risk from hypothermia. To counteract such eventualities, we have recently launched the Lifeline Alarm Appeal. The Lifeline Unit is a sophisticated telephone, easily installed and used basically as a normal telephone — where it differs is that help can be summoned in an emergency either by the emergency button on the telephone itself or by pressing a button worn as a pendant round the neck which activates a radio link with a central control point. A two-way conversation can be held without using the handset. If no verbal response is received at the central control, immediate help is sent. Many local authorities are already using these units. It is the aim of Help the Aged that any elderly person living alone in their own homes (not in sheltered or local authority housing) should have such a Lifeline. It will not replace the caring friend or neighbour, but it can offer the vital reassurance that help will be reached in an emergency.

Last year Help the Aged "spread the word" by distributing 20,000 leaflets called *5 Ways to Keep Warm*. Radio, television and newspapers helped us to get across its important message. It explains about insulation and draught proofing; gives helpful suggestions about paying the bills; what to do if you get cut off; how to deal with a large bill and extra help with fuel bills. It was so successful and sought after that it is being re-issued at the end of this month — this time 100,000 copies will be available for free distribution!

One of the most successful publications for getting information out to elderly people, has been through our newspaper *Yours*. For 12 years we have been producing this each month specifically for older readers. In April of this year it was relaunched in its new format with a new approach to production, distribution and design. By coincidence this month's edition carries a free booklet entitled *Winter Warm Up* giving heaps of tips and advice on how to keep the warmth "in" and the cold "out" over the next few months.

Publicity and awareness are the key words to change. People know that hypothermia is the cause of death in some old people during winter — but appear to think it is some disease or germ that you catch! I wish that newspapers who will surely be reporting some of the hundreds of needless tragic deaths again this winter, wouldn't only print "shock horror headlines", but would do some investigative reporting to find out what contributed to that death. I don't know if this would affect those sitting in warm centrally heated homes, but it might help to increase public awareness and support for a better deal for pensioners.

Until we change society's attitude and force them to recognise the rights of elderly people as citizens — who have contributed and want to go on contributing to the life of this country, to have an adequate income and better housing provision — hypothermia will continue to take its toll. Be assured, Help the Aged will continue to play its part in every way it can, striving at all times to improve the quality of life of elderly people in need.

DISCUSSION

Question: So often the people who need help the most are not
              contacted and in some cases they make sure they are not
              contactable. Many pensioners are not eligible for supple-
              mentary benefit because they have more than the
              approved amount in their savings for a rainy day. They
              are afraid of spending that money and thus exclude
              themselves from the supplementary benefit. Many others
              feel it below their dignity to claim extra allowances. It is a
              well-known fact that between 600,000 and a million
              people would be entitled to some supplementary benefit
              if they claimed.

Answer: *Mrs. Baron*:
              I think I can say we just have to go on publicising it. The
              publication that I mentioned — *Your Rights* — just says
              everything. You have a right to all these benefits and I
              think that we have to keep telling elderly people that it is
              their right, they have worked for it and that it is not a
              charity.

Comment: *Jack Jones*:
              Unfortunately, that is not the case. You can have
              people told of their rights, but rights do not allow them to
              claim if they have so much money in the bank.

Question and The fact that there are about ten million pensioners in
Comment   this country is a tribute to medical science and the
              advances made over many years which have enabled us to
              live longer. One of the medical speakers gave us the

figures for recorded deaths from hypothermia from 1979 to 1983, showing dramatic reductions in recorded deaths from hypothermia. Another speaker mentioned that it is not just the question of money — "money is not the be all and end all". This worries me. I am also worried by the proposed changes in the benefits by the government. One of the proposed changes in housing benefits will force pensioners to meet the first 20 per cent of their rates which they do not meet now. This might account for more than any increase in pensions and other benefits due to be announced shortly. With reduced amount of money in their pockets, the pensioners despite all their attempts, will find it difficult to keep warm. With reduced resources, there will be more pressure on medical and voluntary organisations trying to assist old people. Remembering the thirties, it seems to me that we are rapidly going back to the thirties. I believe that the pensioners, Age Concern, Help the Aged and all other voluntary organisations will have a mammoth task on their hands trying to stem the flood because I believe the recorded incidence of deaths due to hypothermia in a bad winter will be worse than in 1979.

Comment: *Mrs. Rhodes*:
Like Mr. Jones's organisation, we are also worried about the ways of getting through to those people who are on the borderline and who will not claim. Writing any number of pamphlets and books does not seem to help. You have got to get through personally to those people and explain to them that the benefits are not a charity. If you can get through to them, you will have done a great deal towards not only hypothermia, but also the other great illness of the elderly—loneliness.

Question: I am speaking on behalf of people like my mother who is 82, who didn't get a full rise in her pension last year because she had a heating allowance deducted.

Answer:	*Mrs. Baron*: Help the Aged are concerned to find that last year 1½ million pensioners were worse off after deductions than they had been. I feel that it is the duty of the government of the day to look after the rights of the senior citizens. Money plays a vital role in buying heating, better housing conditions, food and better nutrition. It also gives you mobility to go out and visit places of your interest.
Question:	I am surprised to note that a number of medical people did not stay for this part of the conference. Somebody here mentioned about the proposed cuts in various allowances, including housing benefits. I think we ought to be saying that we don't like these changes, and we really have to do something about them.
Comment:	I am a semi-retired doctor and have the greatest sympathy with colleagues who have been here this morning, most of them are hospital consultants. In addition to their clinical responsibilities, they are expected to attend a number of committee meetings for administrative purposes. In addition, a great many of them will be working tonight when the rest of us are at home.
Comment:	*Jack Jones*: I don't think the criticism was aimed at the doctors so much as the feeling that somehow we tend to operate in separate pockets. There isn't enough mixing together and understanding each other. Perhaps the message from this conference, both morning and afternoon sessions, is that there must be more of this sort of meeting organised not only at the Manor House Hospital but elsewhere as well.
	Professor Keatinge: I have learnt a lot and I hope that it has been beneficial to all concerned.

Question: One of the things that worries us very much are the people living in old housing whether they are private or council tenants. The property occupied by the elderly is often run-down. Despite the availability of allowances for draught proofing, etc., the maintenance is not adequate. When talking about the cost of heating, we should not forget about the standing charges which adds £9 or so each quarter to the fuel bill.

Answer: *Jack Jones*:
              The standing charges is not just a matter of the heating costs adding to the gas and electricity bills. The standing charge on telephones is a deterrent for many old people to have a telephone which is a vital life-line to get assistance in emergency, especially for those who live alone.

Comment: We all know that the present level of pensions is hopelessly inadequate. All the conferences and study groups are unlikely to improve this. What we want is more campaigning from all organisations, whether charitable or voluntary, in support of better pensions. Unless we all do something about this, there will be more victims this winter, whether it is a mild winter or a hard winter. With the available money, the elderly often have to choose between adequate food or heating. Lack of either leads to suffering.

Comment: Just a comment about the standing charges. Do you realise that a pensioner loses a week's pension every quarter and a whole month's pension each year on standing charges alone.

Comment: What I want to say has been carefully avoided in a conference like this, but I feel it is necessary to say. Whether we like it or not, politics is one of the keys to the answers to the questions that have been raised here.

Many of the members in my branch are afraid of mentioning politics as the answer. I am not saying that they are afraid of politics as such. They are scared of a certain brand of politics, left wing politics to put it bluntly, because it seems to have all the answers to most of the problems. Some members in my branch of Pensioners Voice treat pensions as pin money because they have another income to rely on. This group of pensioners may be sympathetic to the pensioner problems, but they do not seem willing to do anything about it. They will applaud when somebody proposes to do something, but are unwilling to participate personally. The organisers of any campaign for improvement has to tackle such obstacles.

Comment: I would like to emphasise one point that has not been dealt with so far. It is a fact that a lot has been and is being written telling everybody what ought to be done. It is unfortunate that the majority of the people who need this information cannot read due to illness, old age or some other reason. To overcome this problem, there must be a person-to-person communication. The idea of the community and the good neighbourhood schemes should be actively supported by all concerned.

Summary and Closing Remarks

J. JONES

As outlined in the programme, this Conference started with discussion on the medical aspects of hypothermia and ended with the recognition that hypothermia is an important social problem. As in the case of other illnesses and problems facing the elderly, the management of hypothermia requires active contribution from all sections of the community. The value of this Conference, and I pay tribute to the Manor House Hospital for organising it, is that we have been able to bring together the medical profession, social workers and pensioners. I think we should arrange more of such conferences in the boroughs, in the counties and generally in our society so that a better understanding of the problems of age and the problems of the elderly is obtained. This will help our society to decide what priorities do we allocate to those in need, the children, the disabled and the elderly. In other countries, the elderly and those in need are given a much higher priority than what is given in our society. Whether you call it politics or something else, we have to build up a great deal of public opinion that will change the mind of those in authority so that better provision is made for dealing with the problems of the elderly.

I would like to thank all the speakers for their contributions. It has been a splendid and a most successful conference. I would like to thank all of you who have spent time here today. At this stage of the conference I can recall one of the characters in one of Victor Hugo's books which said "let us build the human society". I feel this is what this conference has been about. I would like to repeat my thanks to all those concerned in organising this conference that enabled the pensioners, doctors and social workers to sit down together and talk about this important problem facing them. I would like to see it repeated on a regular basis everywhere in the country. It will be beneficial to include those in authority. I know that Camden Council has set up a Liaison Committee where pensioners can sit down with the Chairman and others and talk about problems of the elderly in the borough. It helps to understand the problem and to improve the service. I would like to see that happening in every borough. Maybe there will be such occasions when doctors should also be invited to give the benefit of their advice and experience. I appreciate the great importance of campaigning, but we have to campaign on the basis of understanding. We have to campaign with the backing of the wider community, and in that wider community I do count the doctors and social workers as a major factor.

Index